Signals and Communication Technology

For further volumes:
http://www.springer.com/series/4748

Fernando Almeida · Maria Teresa Andrade
Nicola Blefari Melazzi · Richard Walker
Heinrich Hussmann · Iakovos S. Venieris
Editors

Enhancing the Internet with the CONVERGENCE System

An Information-centric Network Coupled with a Standard Middleware

Editors
Fernando Almeida
Maria Teresa Andrade
Telecom and Multimedia Division
INESC Porto Campus da FEUP
Porto
Portugal

Nicola Blefari Melazzi
Dipartimento di Ingegneria Elettronica
Università degli Studi di Roma
  Tor Vergata
Rome
Italy

Richard Walker
Blue Brain Project
Ecole Polytechnique Federale de Lausanne
Lausanne
Switzerland

Heinrich Hussmann
Medieninformatik
Ludwig-Maximilians-Universität
Munich
Germany

Iakovos S. Venieris
School of Electrical and Computer
  Engineering
National Technical University
  of Athens
Athens
Greece

ISSN 1860-4862   ISSN 1860-4870   (electronic)
ISBN 978-1-4471-5372-6   ISBN 978-1-4471-5373-3   (eBook)
DOI 10.1007/978-1-4471-5373-3
Springer London Heidelberg New York Dordrecht

Library of Congress Control Number: 2013953644

Printed on acid-free paper

Springer is part of Springer Science+Business Media (www.springer.com)

Preface

This book describes the main findings and achievements of the CONVERGENCE project, co-funded by the European Union, within the seventh Framework Programme. The project was conceived and submitted to the European Union services in October 2009. It started in June 2010, and ended in May 2013.

The rationale of the project lies in the evolving nature of the Internet. The Internet was originally conceived as an "Internet of Hosts", whose underlying protocols were designed to support exchange of simple unstructured information between well-identified nodes. Today, by contrast, it is becoming an Internet of Things (devices and appliances associated with their own IP address), an Internet of Services (in which users in different localities access different functionalities on different hosts), an Internet of Media (shared and managed across different networks) and an Internet of People (boosted by the explosion of social networking and the emergence of the Web 2.0 paradigm). In this evolving scenario, the key elements are no longer "hosts" but data and services. In other words, what we are observing is a shift from a "host-centric" Internet to a "data-centric" or "content-centric" or "information-centric" Internet.

A new Information-Centric Internet should support not only traditional service models such as client–server and conversational services but also peer-to-peer and publish–subscribe service models. In client–server models, the server provides a function or service to one or many clients, which initiate requests for such services. In peer-to-peer models, each node plays both the role of client and server for other nodes in the network allowing shared access to various resources such as files, peripherals and sensors without the need for a central server. In publish-subscribe models, subscribers register their interest in an event, or a class or a pattern of events, and are subsequently asynchronously notified of events generated by publishers. In practical terms, publishers (i.e. information providers but also simple users) make available in the network content that they produce; subscribers declare their interest for a given content or a pattern or a class of content. Publications and subscriptions are fully decoupled in space but also in time. Normally, subscriptions precede publications, and in this case the subscriber will receive the required content only if and when a publisher will make that content available in the network. In addition, we envisage that subscriptions can also follow publications, and in this case the subscriber will promptly receive the required content as in a traditional request/response or pull model. Subscribers can express interest both for known specific content (e.g. the repair manual for a piece of equipment) or content satisfying a given search criteria (e.g. all songs by a certain musician).

A new Information-Centric Internet should be able to handle any kind of digital data: classical media contents, data about services, but also digital representation of real-world objects (e.g. items of merchandise identified with an RFID) and people (e.g. profiles). Thus, a new Information-Centric Internet should provide structured and flexible data units allowing to contain and transport different kinds of information, each with its own characteristics and needs, together with meta-information, describing the information itself, its characteristics and needs, and possible actions that the network must exert on it.

More specifically, we believe that a new Information-Centric Internet should allow to:

- create a data unit, in a modular and flexible way, defining licenses and rights that apply to the information that it contains;
- sign and/or encrypt a data unit;
- support client/server, conversational, peer-to-peer and publish–subscribe service models; the latter implies the ability to publish a data unit and to subscribe to a data unit, meeting specified criteria;
- search and retrieve a data unit, with the meta-information contained in the data unit easing semantic searches and operation of search engines;
- verify the authenticity of a data unit;
- allow the publisher or owner of a data unit to monitor its use in the network;
- allow to communicate with owners or publishers of a data unit;
- versioning a data unit and linking it to other data units;
- update a published data unit;
- delete a data unit, allowing implementing digital forgetting and garbage collection, deleting 'expired'/obsolete contents from the whole network;
- do all of this in an efficient way, simplifying operations for all the involved players, including providers, business users and final users.

Now, to reach these objectives and provide the functionality and service models listed above, we were faced with two alternatives: (i) design and deploy different applications, each with its own data unit, operating on top of the current Internet in an overlay fashion and adding more and more patches to the basic operation of the network layer; (ii) define a common, standard unit of distribution and transaction; define a middleware implementing complex functionality at the service of all interested applications; transform the current network layer in a new information-centric network.

The project CONVERGENCE adopts the second alternative, providing new results in the following four areas.

CONVERGENCE defines a unit of distribution and transaction, called Versatile Digital Item (VDI), which contains all necessary supporting information, including signalling, control and security/privacy information.

CONVERGENCE defines a middleware supporting sophisticated functional-ities, which we think are too complex to be implemented at the network layer in all routers, namely publish/subscribe services, searching functions, security functions; the data unit of our middleware is the VDI. It is important to observe however that not all CONVERGENCE communications must necessarily use a publish/sub-scribe paradigm. The CONVERGENCE middleware also accepts direct requests to immediately provide specific requested data, with a traditional request–response service model.

CONVERGENCE defines an information-centric network (ICN) layer. Infor-mation Centric Networking (ICN) is a new paradigm in which the network layer provides users with content, instead of providing communication channels between hosts, and is aware of the name (or identifiers) of the contents. The basic functions of an ICN infrastructure are to: (i) address content by adopting an addressing framework based on names, without a reference to the current content location (i.e., location-independent names); (ii) route a user request, based only on the content-name, towards the "closest" location containing the required content; potential locations include not only the origin server of that content but also network caches or even devices of other users that downloaded the same content beforehand; (iii) deliver the content back to the requesting host.

CONVERGENCE defines specific tools and of applications showing the CONVERGENCE potentiality. Tools are reusable Application elements, which facilitate reuse of code in Applications; an Application can make use of several tools. Our tools and applications exploit the VDI concept and make use of our middleware and network functionality, so offering to end-users the advantages brought about by our system. The project designed and implemented four main applications to show the usefulness of CONVERGENCE in four real-life scenarios. The four scenarios are: (i) management of audio–visual material; (ii) management of a large photo archive; (iii) customer relationship management and logistics for the retailing sector, exploiting information about Real World Objects; (iv) augmented lecture podcast service, enabling a collaborative learning environment. Two other applications have been built later on by integrating the four main applications; the first integrated application merges the first (video) and fourth (podcasts) original applications; the second integrated application merges the second (pictures) and third (retail) original applications. The aim of the inte-grated applications is to show that our system is flexible enough to combine different applications in one and to exploit common VDIs.

The end result of the project is the definition and evaluation of a complete ICT system able to provide the following advantages.

Advantages Deriving from Using an Information-Centric Network Layer

1. Efficient content-routing. Even though today's Content Delivery Networks (CDNs) offer efficient mechanisms to route contents, they cannot use network resources in an optimal way, because they operate over-the-top, i.e. without knowledge of the underlying network topology. ICN would let ISPs perform native content routing with improved reliability and scalability of content access. This would be a built-in facility of the network, unlike today's CDNs;
2. In-network caching. Caching enabled today by off-the-shelf HTTP transparent proxies requires performing stateful operations. The burden of a stateful processing makes it very expensive to deploy caches in nodes that handle a large number of user sessions. ICN would significantly improve efficiency, reliability and scalability of caching, especially for video;
3. Simplified support for peer-to-peer like communications, without the need of overlay dedicated systems. Users could obtain desired contents from other users (or from caching nodes) thanks to content-routing and forward-by-name functionality, as it is done today with specialized applications, which, once again, do not have a full knowledge of the network and involve only a subset of possible users;
4. Simplified handling of mobile and multicast communications. As regards handovers, when a user changes point of attachment to the network, she will simply ask the next chunk of the content she is interested in, without the need for storing states; the next chunk could be provided by a different node than the one that would have been used before the handover. Similar considerations apply for multicasting. Several users can request the same content and the network will provide the service, without the need for overlay mechanisms, multicast is an inherent capability in ICN;
5. Content-oriented security model. Securing the content itself, instead of securing the communications channels, allows for a stronger, more flexible and customizable protection of content and of user privacy. In today's network, contents are protected by securing the channel (connection-based security) or the applications (application-based security). ICN would protect information at the source, in a more flexible and robust way than delegating this function to the channel or the applications. In addition, this is a necessary requirement for an ICN: in-network caching requires to embed security information in the content data-unit, because content may arrive from any network or user node and we cannot trust all nodes; thus, end-users must be able to verify the validity of the received data; caching nodes must make the same check, to avoid caching fake contents;
6. Content-oriented quality of service differentiation (and possibly pricing); provision of different performance in terms of both transmission and caching. Network operators (especially mobile ones) are already trying to differentiate quality and priority of content, but they are forced to use deep packet inspection

technologies. ICN would let operators differentiate the quality perceived by different services without complex, high-layer procedures and off-load their networks via caching, a very handy functionality, particularly for mobile operators who can differentiate quality and priority of content transferred over the precious radio real estate;

7. Content-oriented access control, providing access to specific information items as a function of time, place (e.g. country) or profile of user requesting the item. This functionality also allows implementing: (i) digital forgetting, to ensure that content generated at one period in a user's life does not come back to haunt the user later on, and (ii) garbage collection, deleting from the network expired information;

8. Possibility to create, deliver and consume contents in a modular and personalized way; ICN provides opportunities for better customization of the interests of users and the content that is published by providers. This will enable more efficient consumption of content because of better "granularity" in how content is described and identified;

9. Network awareness of transferred content, allowing network operators to better control information and related revenues flows, favouring competition between operators in the inter-domain market and better balancing the equilibrium of power towards over the top players;

10. Support for time/space-decoupled model of communications, simplifying implementations of publish/subscribe service models and allowing "pieces" of network, or sets of devices to operate even when disconnected from the main Internet (e.g. sensors networks, ad hoc networks, vehicle networks, social gatherings, mobile networks on board vehicles, trains, planes). This last point is a very important one, especially to stimulate early take up of ICN in selected (and possibly isolated) environments.

A final overall advantage of ICN, which in a way comprehends the specific advantages listed above, is a simplification of network design, operation and management. Currently, content and service providers have to "patch" shortcomings and deficiencies of IP data delivery by using several "extra-IP" functionalities, such as HTTP proxies, CDNs, multi-homing and intra-domain multicast delivery, to name a few. This implies the involvement of several parties, the use of several specific protocols, the deployment of ad hoc devices and the interplay of different functionalities, often offered and managed by different companies and businesses. Apart from technical complexity, such operations also add management and administrative complexity. In an ICN environment, such diverse functions can be integrated in the network in a smooth and seamless way, e.g. by supporting inherently data replication, caching, multi-homing and multicast delivery.

Advantages for End-Users

CONVERGENCE allows end-users to:

1. Publish VDIs to the CONVERGENCE networks, using the same publish/ subscribe services for different categories of information;
2. Manage and protect the information contained in VDIs, using the same security and privacy standards for all transactions;
3. Define and enforce licenses;
4. Monitor the use of published materials;
5. Communicate with the users of VDIs;
6. Operate on VDIs at all stages in the data life cycle, examine different versions of the same VDI;
7. Update VDIs that have already been published;
8. Delete VDIs that have already been published (digital forgetting);
9. Search for VDIs using semantic search mechanism;
10. Subscribe to VDIs;
11. Define how much identity information they wish to disclose when publishing or searching for information (multiple options available from full anonymity to full disclosure);
12. Enjoy better performance thanks to ICN;
13. Exchange data with nearby connected devices also when disconnected from the main Internet.

Advantages for Developers

1. Exploit the functionality of the CONVERGENCE middleware avoiding the need to use proprietary solutions;
2. Develop basic functionality using the CONVERGENCE tools and focus on adding value to the CONVERGENCE framework.

Advantages for Network Operators

1. Distribute content over the network more efficiently, reducing costs;
2. Differentiate Quality of Service for different categories of content;
3. Exert better control over information transfer and related revenues flows;
4. Compete more effectively with "over the top" players.

This book is organized into ten chapters. The introductory chapter contextualizes the work developed by the CONVERGENCE consortium, analysing the main limitations of the current Internet architecture to support the increasingly stronger user- and content-centricity trends. It provides an overview of the emergent Information Centric Networking (ICN) paradigm, addressing design principles, evolution scenarios and briefly describing the main research initiatives conducted in the last years.

Consequently, the next eight chapters strive to describe to the reader how CONVERGENCE has embraced the quest for answering the above-referred challenges and to contribute with innovative solutions for the future Internet. These chapters provide high-level descriptions of the concepts and design principles behind CONVERGENCE as well as details on how the different components operate individually and how they simultaneously co-operate among them, to deliver the desired functionality whilst facilitating the deployment of advanced business models. The conclusions chapter summarizes the main achievements and contributions of CONVERGENCE towards the future Internet, drawing paths for using and extending the CONVERGENCE outcomes.

Chapter 2 describes the overall architecture of the CONVERGENCE system. The objective of this chapter is to provide a concise, yet complete, understanding of the concepts and high-level functionality of the CONVERGENCE architecture and how its components interact in concrete deployment scenarios.

Chapter 3 covers the design principles, functionality and implementation details of the CONVERGENCE network level, the CoNet. It shows how this Information-Centric Network extends the content-centric paradigm in several aspects, including routing scalability and security handling, among others, whilst providing a graceful incremental solution, backwards compatible with the current Internet.

The Content Level of CONVERGENCE, conceived and implemented as the CONVERGENCE Middleware (CoMid), is described in Chap. 4. This chapter starts by presenting MPEG-M, which is the standard that provides the foundations for CoMid. It then describes the key components of the Content level, which comprise a diversified set of middleware engines to manipulate Versatile Digital Items (VDIs) and extending MPEG-M, as well as semantic tools, namely the Community Dictionary Service (CDS). The chapter describes how the semantic mechanisms are used together with the standardized technology to support the semantic publish–subscribe paradigm. It explains how CONVERGENCE manipulates rich metadata within a content-aware semantic overlay, running on top of the information-centric networking platform, namely on top of the CoNet.

Chapter 5 provides the definition of the Versatile Digital Item (VDI), the basic unit for data distribution used within the CONVERGENCE system. It explains how the VDI extends the scope of the MPEG-21 Digital Item specification to provide a self-contained data package that can be used to encapsulate any kind of digital information in an information centric, publish–subscribe framework.

Security aspects are analysed in Chap. 6. This chapter presents the distributed architecture of the CONVERGENCE security core component (CoSec), describing the use of smart cards as a secure token to provide sensitive security functions on a tamper-resistant device. The chapter starts by introducing the concepts and the architecture of the security infrastructure, then presenting the high-level security functionality, based on a description of the CONVERGENCE basic cryptographic primitives, as well as advanced cryptographic schemes.

Chapter 7 describes the CONVERGENCE's licensing scheme and its governance, using and extending the MPEG-21 part 5 standard to support specific content protection and rights management requirements of the future Internet. It explains how Rights Expression Language data is embedded into the CONVERGENCE data unit, the Versatile Digital Item (VDI) and how digital certificates can be used to enforce the rights and conditions expressed in CONVERGENCE licenses.

The overall CONVERGENCE functionality is explored and benefits are demonstrated, through the use of applications that were specifically designed and developed to implement a set of four use cases, which are likely to become commonplace in the future Internet. Accordingly, Chap. 8 provides a description of those four realistic use cases and the corresponding applications: *Photos in the Cloud and Analyses on the Earth* (under the responsibility of partner Alinari); *Videos in the Cloud and Analyses on the Earth* (under the responsibility of partner FMSH); *Augmented Lecture Podcast* (under the responsibility of partner LMU); and *Smart Retailing* (under the responsibility of partners WIPRO and UTI). This chapter also discusses the impact of the CONVERGENCE technology to the end-users, according to the user feedback obtained from the conducted field trials, as well as further exploitation scenarios of the developed applications. Such additional scenarios were exemplified through an extra use case that was built by integrating three of the CONVERGENCE applications. The integrated scenario proved that CONVERGENCE is flexible enough to combine different applications in one and to exploit common VDIs.

Finally, Chap. 9 presents the envisaged CONVERGENCE business Models for the commercial and non-commercial exploitation of CONVERGENCE applications and technology. This chapter discusses the feasibility and implications of alternative exploitation strategies, identifying competing products and services, as well as market risks and threats.

CONVERGENCE Researchers

CNIT		ALINARI		CEDEO.net	
Nicola	*Blefari Melazzi*	*Andrea*	*De Polo*	*Leonardo*	*Chiariglione*
Giuseppe	*Bianchi*	*Sam*	*Habibi Minelli*	*Riccardo*	*Chiariglione*
Marco	*Bonola*			*Angelo*	*Difino*
Lorenzo	*Bracciale*				
Stefano	*Brogi*				
Matteo	*Cancellieri*	**ICCS**		**FMSH**	
Alberto	*Caponi*	*Angelos*	*Anadiotis*	*Elisabeth*	*De Pablo*
Andrea	*Detti*	*Vasso*	*Giotopoulou*	*Valérie*	*Legrand-Galarza*
Andrea	*Fratini*	*Georgios*	*Lioudakis*	*Francis*	*Lemaitre*
Andrea	*Giancarli*	*Dimitra*	*Kaklamani*	*Manuela*	*Papino*
Sam	*Habibi Minelli*	*Sofia*	*Kapellaki*	*Peter*	*Stockinger*
Leonardo	*Linguaglossa*	*Aziz*	*Mousas*		
Francesco	*Lombardo*	*Charalampos*	*Patrikakis*		
Giacomo	*Morabito*	*Iakovos*	*Venieris*		
Claudio	*Pisa*				
Matteo	*Pomposini*				
Bruno	*Ricci*				
Stefano	*Salsano*	**INESC Porto**		**LMU**	
Salvatore	*Signorello*	*Fernando*	*Almeida*	*Alina*	*Hang*
Dimitri	*Tassetto*	*Maria Teresa*	*Andrade*	*Heinrich*	*Hussmann*
Giuseppe	*Tropea*	*Helder*	*Castro*	*Evgeniva*	*Ivanova*
MORPHO CARDS		**Singular Logic**		**XIWRITE**	
Silke	*Geisen*	*Panagiotis*	*Gkonis*	*Barbara*	*Benincasa*
Thomas	*Huebner*	*Stelios*	*Pantelopoulos*	*Raffaele*	*Di Fuccio*
Andreas	*Kohlos*			*Andrea*	*Pede*

(continued)

(continued)

| Carsten | Rust | | | Richard | Walker |
| Amit | Shrestha | | | | |

UTI		WIPRO			
Bogdan	Ardeleanu	José	Horta		
Ileana-Catinca	Bobric	João	Pereira		
Danut	Clapa	José	Ribas		
Lucian	Corlan	Inês	Santos		
Costin	Ochescu	Daniel	Sequeira		
Mihai	Tanase				

Contents

Chapter 1
Approaches for the Development of Information Centric Networks

Fernando Almeida, Teresa Andrade, Nicola Blefari Melazzi,
Richard Walker, Heinrich Hussmann and Iakovos S. Venieris

Abstract The new generation of applications is following the trend of user-centricity where users are not seen just as consumers, but are active participants, sometimes being the application itself. Likewise, the trend of content or information-centricity is already being adopted, where content and no longer location is being the main driver of network-based operations, more specifically of routing operations. This chapter analyzes the main limitations of current Internet architecture to support such trends and consequently provides an overview of the emergent Information-centric Networking (ICN) paradigm, addressing design principles, evolution scenarios and potentialities. Furthermore, the main research initiatives conducted in the last years concerning the ICN paradigm are presented and discussed.

F. Almeida (✉)
Telecom and Multimedia Division, INESC Porto Campus da FEUP,
4400-103 porto, Portugal
e-mail: almd@fe.up.pt

T. Andrade
INESC TEC, Faculty of Engineering, University of Porto, Porto, Portugal

N. Blefari Melazzi
Electronic Engineering Department, University of Rome "Tor Vergata", Rome, Italy

R. Walker
Blue Brain Project, Ecole Polytechnique Federale de Lausanne, Lausanne, Switzerland

H. Hussmann
University of Munich, Media Informatics Group, Munich, Germany

I. S. Venieris
School of Electrical and Computer Engineering, NTUA, Athens, Greece

F. Almeida et al. (eds.), *Enhancing the Internet with the CONVERGENCE System*,
Signals and Communication Technology, DOI: 10.1007/978-1-4471-5373-3_1,
© Springer-Verlag London 2014

1.1 Introduction

The Internet was originally conceived as an "Internet of Hosts", whose underlying
protocols were designed to support exchange of simple unstructured information
between well-identified nodes. Today, by contrast, it is becoming an Internet of
Things (devices and appliances associated with their own IP address), an Internet
of Services (in which users in different localities access different functionalities on
different hosts), an Internet of Media (media resources shared and managed across
different networks) and an Internet of People (boosted by the explosion of social
networking and the emergence of the Web 2.0 paradigm). In these "new Inter-
nets", the key elements are no longer "hosts" but data and services (or content).

The Networked Society embraces all stakeholders: people, businesses
and society in general (Fig. 1.1). Different stakeholders have different interests and
drivers for adopting ICT solutions. For people, it is more about lifestyle, fun and
"wants" rather than "needs". On the other hand, enterprises are exposed to an
ever-increasing competitive business environment requiring cost reduction,
branding and differentiation. From a society perspective, saving energy, sustainability, efficiency and safety are important drivers.

This interesting paradigm shift, centers the problem on the process of
dissemination of information, through a long set of heterogeneous devices
(e.g., mobile phones, tablets, etc.) with an expected agnostic behavior in terms of
network technology. This approach brings a new demanding challenge in terms of
a well-defined content awareness structure that was initially addressed at the
beginning of the last decade. Typically two main general approaches have been
investigated: (i) content tagging routing, whereby tags are attached to content at
ingress points, describing content attributes which are then used for routing (Mitra
and Maheswaran 2003); and (ii) overlay content-based networks, whereby an
overlay point-to-point network is formed, implementing a high-level routing
scheme (Carzaniga et al. 2004). The subject was also addressed in experimental
terms, within the context of distribution of multimedia resources, by European-
funded research projects such as the Enthrone FP6 project (Carvalho et al. 2007).

Fig. 1.1 Drivers of the
networked society (Höller
and Arkko 2012)

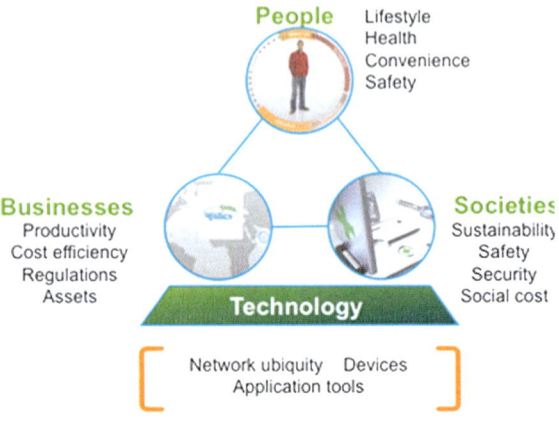

Since then the field has evolved and the currently active and well advanced area of research is known as Content-Aware Networking (CAN), addressing the efficient delivery of multimedia content, where networking equipment manipulates the content to adapt it according to network conditions with the aim of maximizing the quality of experience (QoE) (Ying et al. 2009; Grafl et al. 2012).

Content-centric networking (CCN) or Information-centric networking (ICN), aim at achieving the same goals by radically changing the traditional IP protocol stack, transforming it into chunks of named data. The main concept is based on the use of two types of packets: interest and data packets. Users inject and propagate their interests in the network; network nodes listen to these interest packets, responding with data packets if they have the required data (Van Jacobson et al. 2009). It essentially tries to deal with the lack of a global content naming and addressing scheme, as well as with the need for unified content access and consumption. These current limitations also restrict the capability of the networks to seamlessly adapt content and offer the appropriate QoE to the end users.

The proposed benefits of ICN are wide. Within such a system, content would be uniquely addressed with the ability to be accessed from any location using a simple request/reply paradigm. According to Tyson (2010), this would make application development simpler and allow an array of network-level optimizations to be performed relating to the storage and distribution of content. While imaginative solutions through incremental changes to the network stack or overlay networks have been successfully proposed (Broberg et al. 2009; Demmer et al. 2009; Polyzos 2011), it is widely admitted in literature that these solutions have also limitations and drawbacks, in terms of scalability, security, mobility, and manageability, to encourage the creation or deployment of more ambitious and innovative services (Smetters and Jacobson 2009; Carofiglio et al. 2011; Perino and Varvello 2011).

1.2 Current Internet Limitations

The Internet architecture is a constantly and rapidly evolving interconnection of thousand of networks that act as simple carriers providing basic packet delivery services without guarantees and based on locations. This means that they make their best effort to try to deliver to receivers anything that senders want to send using IP addresses to identify end-hosts for data forwarding, being unaware of the content that is being delivered.

The current Internet architecture and its main supported functionalities are illustrated in Fig. 1.2.

The Internet architecture consists of the following types of nodes:

- Content servers or caches (either professional or user-generated content and services);
- Centralized, decentralized or clustered servers, including search engines and supporting servers;

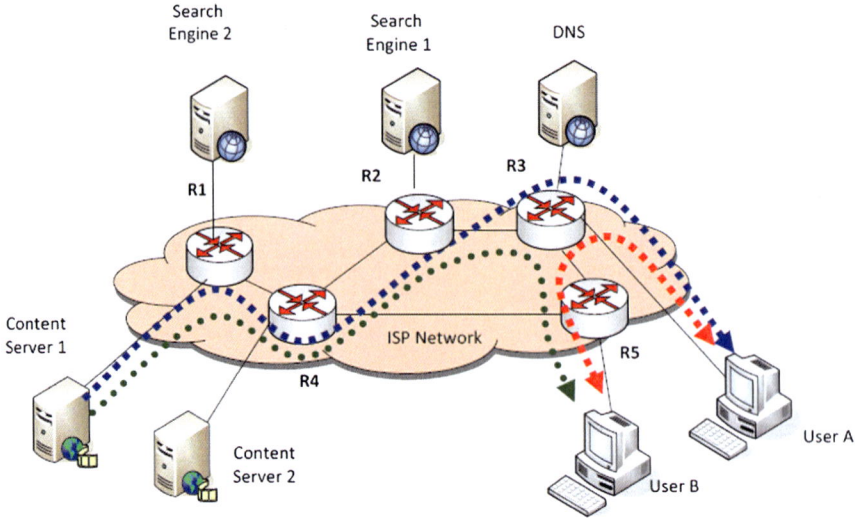

Fig. 1.2 Today's network architecture (FIA 2009)

- Core and edges routers and residential gateways;
- Users connected via fixed, wireless or mobile terminals.

The above architecture works for current applications and usage, and will continue to work adequately if there is sufficient resource (e.g., bandwidth capacity, low delay, low latency, etc.) in the system do deliver it. However, the same architecture would not be suitable if there are billions of connected devices; users demand high resolution videos that require increased bandwidths or more and more users conduct delay-critical real time video and audio communications over the Internet. Additionally, if we shift into a wireless scenario, taking into account the massive growth of mobile devices, the bottleneck can be considerable. Supporting these requirements efficiently in an integrated wired-wireless global network will require major changes.

Actually, the vast majority of today's traffic is indeed driven by information production and retrieval, ranging from simple RSS feed aggregators to advanced multimedia streaming services, including user-generated or dynamic Web content. The relationships between network entities are not restricted to the view of network topology, but also represent social or content-aware connections between users that can additionally share common interests (e.g., newsgroups, online photo, video sharing and social networks) (Moustafa and Zeadally 2012). Therefore, the resulting graph modeling today's Internet is becoming very complex. Nodes are mainly pieces of content, rather than endpoints (locators) addressed by the underlying IP protocol, and are connected by one or more type of interdependency based on the notion of intention, interest or policy-based membership (FIA 2009). While still overlaying on top of the host-to-host conversation model, all the current workarounds for Internet support of emerging content-centric applications increase the complexity and do not efficiently map all the relevant ties between the nodes in terms of security, mobility, multicast, scalability and quality of service (QoS) (Jian 2006; Feldmann 2007; Roscoe 2006).

These required changes will only be supported in the Internet through massive investment, and even so, the resulting architecture may contain unstable characteristics. As a consequence, it will be necessary to consider an evolution of the Internet architecture based on the concept of information-centric networks, which will lead to much more efficient use of available resources and provide a business environment that encourages investment.

1.3 Information Centric Network

1.3.1 Design Principle

Traditional communication networks have always been based around establishing communications paths to specified end-points, connected to servers that would in turn contain content, media and services. This model requires users to know where content is located or even the methods through which the content should be accessed.

However, this model is no longer appropriate for advanced Networked Media Systems. The location of content, servers, services or users should not be an issue of concern. One of the major research challenges is, therefore, how to make communications network as location-independent as possible. Furthermore, as applications become more distributed and interactive, alongside increasing content demand, many of the traditional network functions of naming, addressing, routing and forwarding become strongly influenced by the applications and the content/media in question.

In an attempt to answer these challenges and optimize content delivery to end-users, a number of content distributing technologies have emerged in the past years. Such is the case of content distribution networks (CDNs) and Peer-to-Peer (P2P) systems. These solutions enable the access to content based on its name, regardless of the location of the server or peer that is actually providing the requested content. However, these solutions employ different means to achieve the goal of optimal content delivery and location-independence. Accordingly, deployed solutions act as separate networks, which means that it is not possible to access named content abstracting the underlying distribution infrastructure. Furthermore, although they allow users to access content using its name, without the need to know the location of the specific sources of that content, they still use the underlying networking functionality based on routing-by-location.

ICN, on the other hand, is a longer-term approach to respond to the current Internet requirements, aiming at incorporating right at its core functionality to support the routing-by-name paradigm. Nonetheless, even though not adopting intrinsically the routing-by-name concept, the CDN and P2P solutions already in use constitute the first approaches towards a named-centric Internet and will thus play an important role in optimizing content distribution in the immediate future.

Content Delivery Networks (CDNs) are privately owned networks, deployed as an overlay, with the goal of optimizing the delivery of content to end-users in terms of performance, availability and cost. They contribute to improving the user QoE without requiring Content Service Providers (CSP) to bear additional major costs. QoE is improved via three main features, achieved essentially through the use of multiple caches (known as surrogates) placed in strategic locations: lower latency, higher throughput and increased robustness/availability. Containing costs is achieved through lower bandwidth usage in the core network (due to the cached content) and the requirements of internal infrastructures (such as repository/storage capacity and space). Users are not aware of the cache server that is actually delivering them the content. They simply request the desired content using a content locator (such as a URL). The CDN management system is then responsible for selecting the surrogate that will optimize the delivery of content. This process thus decouples the content identifier from the actual location of the content.

A Content Distribution Network (CDN) is a networked system of interconnected computers that cooperate to deliver efficient and predictable distribution of contents. Its operation is transparent to the users and it is essentially based on geographic location: content delivery servers, replicating the original content, are placed as near as possible to communities of users, thus creating the illusion of proximity between origin content and end-users. This is one solution to the growing problems of distributing contents effectively over the Internet, as bandwidth requirements increase (with the exponential rise of available content as well as of user demand) and the QoE requirements become stricter.

CDNs create copies of the content (either on demand or preemptively) to strategically placed servers hosted by CDN providers, called surrogate servers, and deliver it to end-users based on their characteristics (Buyya et al. 2008). CDN providers are either commercial (i.e. Akamai, EdgeCast, Limelight) or academic/free (i.e. Coral, CoDeeN), and sign contracts with Content Providers. Thus, when a user requests some content from a Content Provider, the requests are automatically redirected to the closest/more adequate CDN sever. The scheme of the CDN structure is shown in Fig. 1.3.

The CDN architecture consists of a content delivery infrastructure to host multiple copies of content, a routing infrastructure so that users' requests are redirected from the Web to CDN servers and copies of content are kept up-to-date, a content distribution infrastructure responsible for moving sources of content close to requesting users, and an accounting infrastructure which accounts users' accesses to CDNs and records CDN servers' traffic and usage, in order to support CDN provider's billing operations.

In general, there are two overall architectures that a CDN provider needs to choose from in order to structure their system; the active network approach and the overlay approach. The former requires the use of network components (i.e. routers and switches), which are not only used to forward packets but also, with the use of special software, to identify application types and use custom policies for each

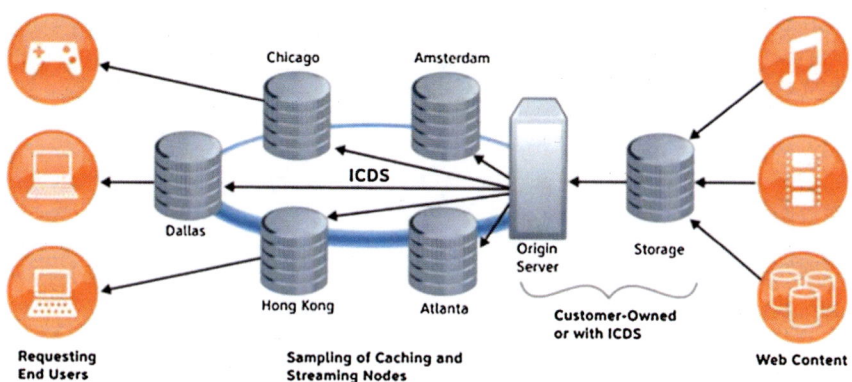

Fig. 1.3 CDN Scheme in an intelligent content distribution service (Salguero 2010)

to route the content. The latter uses application-specific caches and servers to distribute a set of specific content types (static web pages, live TV etc.). The active network approach actively uses network components to assist the delivery of content whereas the overlay approach does not actively make use of them other than for providing basic network connectivity. Some CDN providers may use the two approaches in combination (Sjöberg 2008).

Another important aspect of the CDN infrastructure is the way that servers are placed geographically to optimize content delivery. The placement is decided using algorithms to calculate the most efficient locations (Pallis 2006). Once the placement is decided, CDNs make use of content replication and caching techniques in order to distribute the content closer to the potential end users, rather than repeatedly transmitting identical versions of the content from an origin server. Since surrogate servers are not meant to serve as exact replicas of the origin server, due to their capacity limitations, only a subset of the content is usually distributed to them. Different distribution approaches are used for different circumstances and they can affect the efficiency of the end-to-end content transfer (Pathan 2007). The three most popular include cooperate push-based, non-cooperative pull-based and cooperative pull-based.

Since the users require a global identifier to request each piece of content, CDNs need a mechanism to route their request to the closest and appropriate surrogate server that can more efficiently serve them. This redirection can be done either via DNS, where DNS servers redirect users' requests to optimal CDN servers, based on users' characteristics (users' location, network topology and condition, servers' health and load) or by URL rewriting, where requested URL links are rewritten and requests are redirected to optimal CDN servers. With these techniques, CDNs improve the load of the servers and the latency of the distributed content consumption.

ICN shares the optimization goals of CDNs to deliver content. Apart from the fact that CDNs constitute an already deployed solution, whereas ICNs are still under research, one of the main differences between CDNs and ICNs is that the former are always implemented as an overlay, on top of existing network infrastructures (which route by location). By contrast, ICNs are usually implemented at the network level, using ICN-enabled equipment that is capable to route-by-name. Additionally, the tools they employ to operate and achieve optimization may differ. ICNs normally rely on network elements (mostly routers) that are able to cache content and to support QoS as well as traffic differentiation. There is no central management in ICNs, with requests being routed throughout the network, based on a content identifier. Routers with caching capability are dynamically populated with content based on actual user-demand. There is thus no prior planning on how to place content on the network. In CDNs, content is previously distributed to servers placed in selected locations (the surrogates, at the network edges), according to predicted user-demand. There is a centralized management system in charge of redirecting requests to specific surrogates, according to some optimization criteria. A requested content with a given identifier may be re-directed to surrogates placed in different locations, depending on the origin of the request and on the optimization criteria. Even though both ICNs and CDNs decouple content identifiers from location, in ICNs this is an intrinsic/built-in aspect, whereas in CDNs it is a consequence of their deployment topology and optimization goals.

It is foreseen that the reach or "footprint extensions" of ICNs (covering larger areas and audiences) will be dynamically increased as user-demand grows and new ICN-enabled routers are deployed. For CDNs to enlarge their area of intervention, there are two possibilities: (1) the CDN provider re-structures his surrogate deployment architecture alongside the management system; this implies a re-planning of the distribution of content across surrogates; (2) different CDN providers interconnect their infrastructures. Either way, it is a decision up to the CDN providers, although the second option requires minimal modifications to existing infrastructures.

An example of benefits that could arise from the interconnection of CDNs can be illustrated by a simple example[1] in which a Content Service Provider 1 (CSP1) has an agreement established with a CDN Provider (CDNPx) for the delivery of its content; at some point in time, CDNPx agrees with a second CDN Provider (CDNPy) to interconnect their CDNs. When a given customer of CSP1 requests content, CDNPx may choose to deliver the content using its own infrastructure or using the infrastructure of CDNPy. If, for instance, CDNy is an access network to which CSP1's customer is attached to, the best option would be to deliver the content via CDNy. Accordingly, CDNPx can redirect the request to CDNPy and the content is actually delivered to the user by CDNPy. With this example the following benefits can be enumerated:

(1) Benefits to the user: better QoE, because the content is delivered from a nearby surrogate, thus achieving lower latency, and avoiding bottlenecks;

---

[1] Based on an example provided in IETF RFC 6770 available at http://tools.ietf.org/html/rfc6770.

(2) Benefits to CDNPx: without the need to deploy new and extensive network infrastructure, it may offer an enhanced service to the Content Service Provider (and to the user ultimately), thus potentially increasing its revenue;

(3) Benefits to CDNPy: without having a direct contract with CSP1, it may receive some compensation for the delivery of its content;

(4) Benefits to CSP1: it only needs to establish a contract to one single CDNP whilst being able to offer a QoS to its customers as it would have established contracts to multiple CDN providers.

Despite the potential benefits of interconnecting CDNs, notably that of "footprint" extension, most of existing CDNs are still stand-alone networks (IETF 2012). However, this state of things is expected to change in the very near future, especially fostered by work that is being conducted within the IETF, ETSI, EFIA, Eurescom and other organizations interested in the evolution of Internet towards the ICN paradigm.

1.3.2 Benefits

Information Centric Networks brings the following main advantages (see e.g., Blefari Melazzi et al. 2013):

1. Efficient content-routing. Even though today's CDNs offer efficient mechanisms to route contents, they cannot use network resources in an optimal way, because they operate over-the-top, i.e. without knowledge of the underlying network topology. ICN would let ISPs perform native content routing with improved reliability and scalability of content access. This would be a built-in facility of the network, unlike today's CDNs;

2. In-network caching. Caching enabled today by off-the-shelf HTTP transparent proxies requires performing stateful operations. The burden of a stateful processing makes it very expensive to deploy caches in nodes that handle a large number of user sessions. ICN would significantly improve efficiency, reliability and scalability of caching, especially for video (Detti et al. 2012);

3. Simplified support for peer-to-peer like communications, without the need of overlay dedicated systems. Users could obtain desired contents from other users (or from caching nodes) thanks to content-routing and forward-by-name functionality, as it is done today with specialized applications, which, once again, do not have a full knowledge of the network and involve only a subset of possible users;

4. Simplified handling of mobile and multicast communications. As regards handovers, when a user changes point of attachment to the network, she will simply ask the next chunk of the content she is interested in, without the need of storing states; the next chunk could be provided by a different node than the one that it would have been used before the handover. Similar considerations apply for multicasting. Several users can request the same content and the

network will provide the service, without the need of overlay mechanisms, multicast is an inherent capability in ICN.

5. Content-oriented security model. Securing the content itself, instead of securing the communications channels, allows for a stronger, more flexible and customizable protection of content and of user privacy. In today's network, contents are protected by securing the channel (connection-based security) or the applications (application-based security). ICN would protect information at the source, in a more flexible and robust way than delegating this function to the channel or the applications (Van Jacobson et al. 2009). In addition, this is a necessary requirement for an ICN: in-network caching requires to embed security information in the content data-unit, because content may arrive from any network or user node and we cannot trust all nodes; thus, end-users must be able to verify the validity of the received data; caching nodes must make the same check, to avoid caching fake contents;

6. Content-oriented QoS differentiation (and possibly pricing); provision of different performance in terms of both transmission and caching. Network operators (especially mobile ones) are already trying to differentiate quality and priority of content, but they are forced to use deep packet inspection technologies. ICN would let operators differentiate the quality perceived by different services without complex, high-layer procedures (Oueslati et al. 2011), and off-load their networks via caching, a very handy functionality, par ticularly for mobile operators who can differentiate quality and priority of content transferred over the precious radio real estate;

7. Content-oriented access control, providing access to specific information items as a function of time, place (e.g., country), or profile of user requesting the item. This functionality also allows implementing: (i) digital forgetting, to ensure that content generated at one period in a user's life does not come back to haunt the user later on, (ii) and garbage collection, deleting from the network expired information;

8. Possibility to create, deliver and consume contents in a modular and personalized way; ICN provides opportunities for better customization of the interests of users and the content that is published by providers (SKA 2012). This will enable more efficient consumption of content because of better "granularity" in how content is described and identified;

9. Network awareness of transferred content, allowing network operators to better control information and related revenues flows, favoring competition between operators in the inter-domain market and better balancing the equilibrium of power towards over the top players;

10. Support for time/space-decoupled model of communications, simplifying implementations of publish/subscribe service models and allowing "pieces" of network, or sets of devices to operate even when disconnected from the main Internet (e.g., sensors networks, ad-hoc networks, vehicle networks, social gatherings, mobile networks on board vehicles, trains, planes). This last point is a very important one, especially to stimulate early take up of ICN in selected (and possibly isolated) environments.

A final overall advantage of ICN, which in a way comprehends the specific advantages listed above, is a simplification of network design, operation and management. Currently, content and service providers have to "patch" shortcomings and deficiencies of IP data delivery by using several "extra-IP" functionalities, such as HTTP proxies, CDNs, multi-homing and intra-domain multicast delivery, to name a few. This implies the involvement of several parties, the use of several specific protocols, the deployment of ad hoc devices and the interplay of different functionalities, often offered and managed by different companies and businesses. Apart from technical complexity, such operations also add management and administrative complexity. In an ICN environment, such diverse functions can be integrated in the network in a smooth and seamless way, e.g., by supporting inherently data replication, caching, multi-homing and multicast delivery.

On the cons side, ICN has some drawbacks and challenges. A first, obvious con, is that it requires changes in the basic network operation. A second con is that it raises scalability concerns: (i) the number of different contents and corresponding names is much bigger than the number of host addresses; this has implications on the size of routing tables and on the complexity of lookup functions; (ii) in some proposed ICN architectures (Van Jacobson et al. 2009), delivering contents back to requesting users requires maintaining states in network nodes.

1.4 Research Initiatives

A range of different research initiatives and solutions have been proposed to mitigate some of today's problems, as highlighted in the previous sections. Some of such initiatives, briefly described in this section, include caching and load-balancing techniques in CDNs, dynamic increase of infrastructure, adaptation of multimedia contents and better QoS and QoE. Various research initiatives are investigating alternative approaches, including native information-centric network technologies, dynamic adaptation of content and optimization of overlay networks. The goal of these initiatives tries to essentially looking for the following answers:

- Should content access, storage and delivery be decoupled from the network layer or are there significant advantages in a more tightly coupled content-aware network infrastructure?
- Do the advantages offered by an information-centric network layer outweigh the costs of deploying a new infrastructure? Are there hybrid solutions that achieve the performance and usability gains without impacting today's network? Do we need a new clean state approach or is current IPv4/IPv6 sufficient?
- To what degree should applications and overlay networks cooperate with underlying networks to optimize content delivery, reducing infrastructure cost and improving QoE for the user?

- How should content be pushed closer to end-users to reduce inter-domain traffic and other delivery costs? How can CDNs interwork and cooperate with one other to balance load and extend geographic reachability?

Many ICN architectures have been proposed so far in the framework of research projects, including Publish-Subscribe Internet Technology (Ahlgren 2012). Such is the case of the European-funded research projects PURSUIT, 4WARD, COAST, SAIL, DONA, COMET and CONVERGENCE. Additionally, several initiatives have emerged within the context of standardization organizations, notably the Internet Research Task Force (IRTF) with goals similar to IETF but focusing on longer-term research issues related with the Internet; the European Future Internet Alliance (EFIA); the European Telecommunications Standards Institute (ETSI); among others.

In next sections we will analyze the main concepts behind these architectures.

1.4.1 DONA

Data-Oriented Network Architecture represents a new paradigm that suggests that current Internet limitations reside essentially in how Internet names are structured and resolved (Koponen et al. 2007). Therefore, it proposes replacing DNS names with flat, self-certifying names, and replacing DNS name resolution with a name-based anycast primitive that lives above the IP layer.

DONA improves data retrieval and service access by providing stronger and more architecturally coherent support for persistence, availability, and authentication. It can also be extended to provide support for caching and RSS-like updates. However, DONA's impact is not limited to data and service access; they use these applications as motivating examples because they force us to think differently about some fundamental issues, but most of these issues are not particular to data/service access. As a result, as we describe below, DONA's overall design has architectural implications that range far beyond data/service access. DONA's name-based anycast primitive is useful for many kinds of resource discovery; for instance, it can provide the basic primitives underlying SIP (Session Initiation Protocol), support host mobility and multi-homing, and establish forwarding state for inter-domain multicast. Placing anycast at the naming layer, rather than at the IP layer, allows us to design for functionality rather than be limited by concerns about scalability, since the mechanisms need not operate at link speed.

There is another issue where historical design is at odds with current usage. The original Internet architecture, following the end-to-end principle, intended the network to be a purely transparent carrier of packets. Today, however, the various network stakeholders (such as enterprises) use middleboxes to improve security (e.g., firewalls, proxies) and accelerate applications (e.g., caches) (Blumenthal and Clark 2001). Because DONA's anycast name resolution process follows

Table 1.1 Comparing DONA and CCN architectures (Koponen et al. 2007)

	CCN	DONA
Naming	Structured, human-readable	Unstructured, self-certifying name carry a hash of the public key
Public key cryptography	PKI associates a key with name; names and content are signed	A directory maps public key to real-world identity, content is signed
Name governance	Centralized or distributed	Distributed
Name resolution	Integrated with routing	DHT resolution to Internet address
Routing	Routes hierarchical names	Uses internet routing for data
Caching	Integrated with routing	Can route through caches
Layering	Independent of IP and other layers	Shim between IP and transport

essentially the same administrative path as the ensuing data packets, DONA can treat the stakeholders along the path as relevant Internet actors. This allows DONA to provide clean support for network-imposed middleboxes. This is not a repudiation of the end-to-end principle, in that functionality is still provided at the end points; it is merely a recognition that operators should have, at their disposal, architecturally coherent mechanisms to control how and what traffic traverses their network.

DONA thus constitutes an approach to ICN, which works differently from the CCN approach originally proposed by Jacobson (Van Jacobson et al. 2009), as it can be see by the information contained in Table 1.1.

1.4.2 PURSUIT

The PURSUIT initiative contributes to the larger vision of Internet by focusing on changing the routing and forwarding fabric of the global inter-network so as to operate entirely based on the notion of information (associated with a notion of labels to operate the fabric on) and its surrounding concerns, explicitly defining the scope of the information. While the project does not embed the higher-level semantics of information into the network, it intends to devise means that will enable the higher levels to embed concerns and social structures surrounding this information deeply within the architecture (PURSUIT 2012).

PURSUIT formulates the objectives as designing, developing and evaluating a novel information-centric pub/sub-based inter-networking architecture that:

- Provides an improved impedance match towards application-level concepts: One of the major functions of an inter-networking architecture is to provide an appropriate interface to network functions that be effectively utilized by application developers;

- Provides tussle delineation of crucial functions: Building a distributed system requires the mediation of interests stemming from (distributed) stakeholders. The occurring tussles (i.e. the conflicts of such interests) need to be accommodated by the architecture in order to provide a suitable ground for innovation and overall societal viability;
- Enables optimization of sub-architectures: The inter-networking architecture provides a unifying approach for inter-networking disparate systems on global scale. PURSUIT assumes that the current heterogeneity of technological approaches to networking in particular domains will prevail, expressed in a variety of wireless and wireline technologies. It is the clear goal of each infrastructure provider to optimally utilize the resources its infrastructure provides. Hence, optimization of individual sub-architectures is an important design goal in order to make the inter-networking architecture viable in large scale;
- Provides high performance: The inter-networking function is crucial to make distributed applications work in high performance. Hence, developing design choices that follow suit with the expected evolution of technologies such as optical or new wireless is essential for any adoption of the network architecture;
- Scales to the needs of the Future Internet: Scale is essential for the inter-networking function in a global Internet. The proliferation of digital technologies drives the need for scale in the future, driven by increasing digitization of material that is being made available in the network.

PURSUIT, similar to its predecessor publish-subscribe internet routing paradigm (PSIRP), is an architecture design project that attempts to build solutions for a new form of internetworking. The project intends to undergo all the stages from early evaluation of existing work to design of a solution that is intended to be very different from what we know today in our networks. The PSIRP project established and successfully tested a design methodology that has proven to be effective in a project of the size of PSIRP (and PURSUIT) as well as accommodate the desire of individual researchers to perform world-class research. This approach has been also adopted for PURSUIT.

The major challenge in PURSUIT is to create the right combination of bottom-up ideas and top-down rationalization (which can be seen as forming a somewhat rigorous design process). The promoters of the initiative recognize that the bottom-up approach is of utmost importance to push the limits within the goals of the PURSUIT project. The team structure of PURSUIT encourages this bottom-up approach and it is expected that in particular design choices are pushed forward in this manner.

This view on PURSUIT approach, as a combination of bottom–up work and top–down rationalization, leads to the following methodology for the architecture design (Fig. 1.4).

There are many scenarios in which the development of a PURSUIT Internet would be beneficial, ranging from content distribution over improvements to the World Wide Web and collaborative value chains to sensor networks. Within the work in the predecessor project PSIRP, PURSUIT developed a set of specific

Fig. 1.4 Architecture design principles of (PURSUIT 2012)

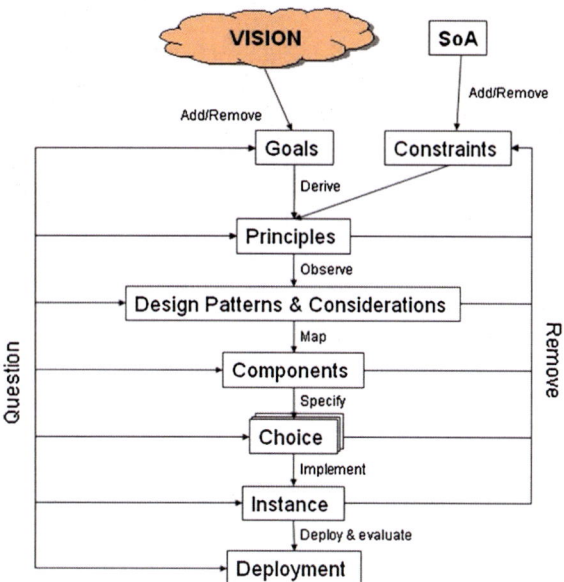

examples that intends to highlight the benefits of an information-centric Internet in terms of (i) information access anywhere and anytime, (ii) security and safety, and (iii) policy-based handling of information.

1.4.3 4WARD

4WARD has performed research on the architecture of a Future Internet adopting a "clean slate" research approach. This means that the practical constraints of evolving from the existing TCP/IP-based network architecture is temporarily ignored in the interest of discovering a design that is ideally adapted to present and expected future usage and is not forced to adapt to architectural decisions made some 30 years ago with quite different objectives and constraints. An architecture following this approach may be seen as a target for the current network to evolve to. It may alternatively be seen as the blueprint of a parallel architecture that could coexist and interoperate with IP, gradually expanding and taking over the functions of the old network (4WARD 2012).

The strategic objective of 4WARD is to increase the competitiveness of the European networking industry and to improve the quality of life for European citizens by creating a family of dependable and interoperable networks providing direct and ubiquitous access to information. 4WARD's goal is to make the development of networks and networked applications faster and easier, leading to both more advanced and more affordable communication services.

4WARD approach combines on one hand innovations needed to improve the operation of any single network architecture and on the other hand multiple different and specialized network architectures that are made to work together in an overall framework. More specifically, 4WARD works:

- On innovations overcoming the shortcomings of current communication networks like the Internet;
- In a framework that allows the coexistence, inter-operability, and complementarity of several network architectures;
- In an integrated fashion, avoiding pitfalls like the current Internet's "patch on a patch" approach.

4WARD introduces a New Architecture Concepts and Principles (NewAPC) and explores a new approach to allow for a plurality and multitude of network architectures: the best network for each task, each device, each customer, and each technology. Unlike the multitude we had in the past, where different incompatible technologies were competing with each other, NewAPC is working on developing an architecture framework that will allow networks to bloom as a family of interoperable networks. Networks will coexist and complement each other, each of them addressing individual requirements such as mobility, QoS, security, resilience, wireless transport and energy-awareness.

One of the basic tenets of 4WARD is that the Future Internet shall allow multiple networking solutions to coexist, not only in the link and the application layer as in the Internet today, but also in the network and transport layers. Network Virtualization is ideally suited to allow the coexistence of different network architectures, legacy systems included. Virtualization is thus not only an enabler for the coexistence of multiple, possibly revolutionary, architectures, but also provides a smooth path for the migration towards more evolutionary approaches. This way, virtualization can help to keep the Internet evolvable and innovation-friendly, particularly since it can mitigate the need to create broad consensus regarding the deployment of new technologies among the multitude of stakeholders that make up today's Internet. By decoupling the infrastructure from the services, virtualization can provide the opportunity to roll out new architectures, protocols, and services without going through the slow and difficult process of creating such consensus.

Virtualization further provides a general approach for network service providers to share a common physical infrastructure. This is particularly beneficial in network domains where the deployment costs per user are predominant and are an encumbrance for frequent technology replacement, as is the case for instance in access networks. The architecture adopted by 4WARD for virtualization management is provided in Fig. 1.5.

4WARD introduces the In-Network Management, which is a novel paradigm to manage networks in the Future Internet. Discovery of network capabilities and adaptation of management operation to current working conditions are key elements in the novel management paradigm.

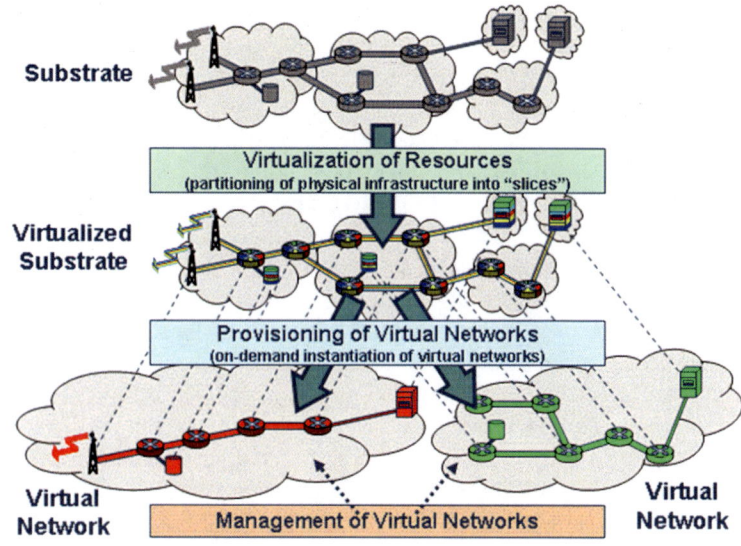

Fig. 1.5 Virtualization management architecture (4WARD 2012)

In traditional Internet management (Fig. 1.6), the management functionality resides outside the network, in dedicated management stations and servers. In commercial networks, interactions between these elements often occur out-of-band, through special communication networks. For emerging large-scale, dynamic

Fig. 1.6 Traditional internet
management approach
(4WARD 2012)

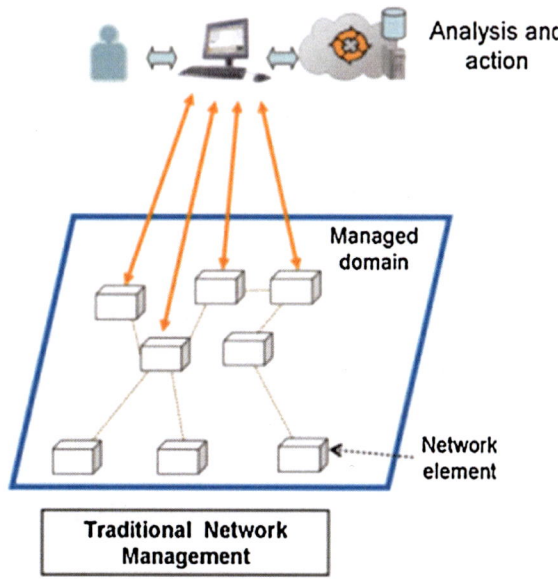

network environments however, the approach turned out to be inadequate and alternative approaches must be developed.

In-Network Management (Fig. 1.7) is a new paradigm for network management, where management functions come as embedded capabilities of the devices. With this approach, network elements have embedded "default-on" management capabilities, consisting of several autonomous components that interact with each other in the same device and with components in neighboring devices. Glued together with a set of discovery and self-organizing algorithms, the network elements form a thin "management plane" embedded in the network itself.

The In-Network Management paradigm can be interpreted as pushing management intelligence into the network, and, as a consequence, making the network more intelligent: as a consequence, objectives and costs of management operations can be adapted according to local working conditions. The network, which now includes the management plane as a part, can execute end-to-end management functions on its own and perform, for instance, reconfigurations in an autonomous fashion. It reports results of management actions to an external management system, and it triggers alarms if intervention from outside is needed.

The architecture of In-Network management first of all models how management capabilities are embedded inside the services of a node. On this basis, it is then possible to compose them, in such a way that the embedded functions are coordinated with each other. Out of smaller autonomous components, more complex management functions can be constructed in the management plane.

The goal of In-Network Management is to achieve scalable, robust management systems with low complexity for large-scale, dynamic network environments. The guiding principles to achieve this goal are decentralization and self-organization.

Fig. 1.7 In-network management approach (4WARD 2012)

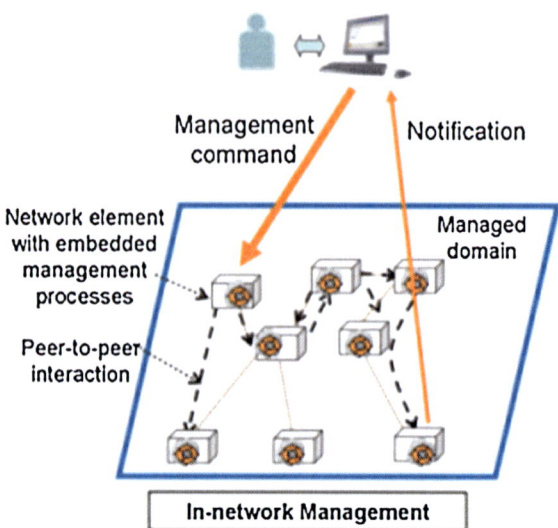

The networking of information approach was also an important contribution of 4WARD. The traditional role of networking has been to interconnect remotely located devices like computers or telephones. This function is increasingly recognized to be ill-adapted and inadequate for the information-centric applications that currently generate the vast majority of Internet traffic.

In 4WARD a different approach was taken. Instead of the node-centric paradigm, the project adopts an information-centric paradigm. In this paradigm, the communication abstraction presented to applications is based on transfer of application data objects instead of the end-to-end reliable byte-stream used by the majority of applications today.

The current semantic overload of the IP-address as both node identifier and locator, indicating the current point of attachment in the network topology, is replaced by a clear separation of information self-certifying object identifiers and locators. Several models for abstracting the location and focusing on networking between (mobile) hosts have been proposed, e.g., the Host Identity Protocol (HIP), the Internet Indirection Infrastructure (I3), the Layered Naming Architecture and the NodeID proposal. 4WARD will build on this prior work and by taking it one step further, the user will be able to design a networking architecture where mobility, multi-homing and security is an intrinsic part of the network architecture rather than add-on solutions. It will also allow users to gain increased control over incoming traffic, which will enable new possibilities for defending against denial of service attacks. The self-securing property also intrinsically facilitates possibilities for effective content protection and access rights management.

The need for ICN is manifested by the increasing number of overlays that are created for the purpose of information dissemination (e.g., Akamai CDN, BitTorrent, Skype, and Joost). Their objective is to distribute information by relying on users to exchange pieces of data between themselves, massively distributing the load away from any central server, and scaling automatically to any group size. 4WARD will integrate much of the functionality of these overlays, including caching functions where the 'copies' are as the originals, in a common and open information networking service that can generally be used by applications.

4WARD extends the networking of information concept beyond "traditional" information objects (e.g., web pages, music/movie files, streaming media) to conversational services like telephony, and store-and-forward services like email. Special attention will be paid to how this affects wireless communication and to how services can be made to work in an environment with a heterogeneous and disruptive communication infrastructure.

1.4.4 COAST

The COAST project aims to build Future Content-Centric Network (FCCN) overlay architecture able to intelligently and efficiently link billions of content sources to billions of content consumers and offer fast content-aware retrieval, delivery and streaming, while meeting network-wide Service Level Agreements (SLAs) in content and services consumption (COAST 2012). This will be achieved by combining intelligent network caching, searching, and network, terminal and user context awareness.

In short, we may specify the COAST functionality as a FCCN that:

1. Provide mechanisms to get the location of contents;
2. Identify/analyze what content and traffic is flowing through the network routers;
3. Replicate and cache the content efficiently at the "best" place in the network;
4. Dynamically identify what is the "best" host/cache and the end-to-end path (in terms of both efficiency and network-friendliness) for content delivery and streaming to a user;
5. Provide the "best" Perceived Quality of Service (PQoS) to the user by inter-actively adapting the content, based on the user and the terminal capabilities, requirements and context.

A major key enablers of COAST is to allow for incremental deployment of intelligent services into the network. COAST will progressively move intelligence in the network by implementing new networking nodes (edge routers, home gateways, terminal devices), with visibility into the type of data that they are carrying or caching. This content-level information can then be used by the routers to make more intelligent retrieval, routing and data handling decisions. Unlike other approaches, COAST relies on distributed "on-line" searching of published content (digital objects, streams and services) and "on-the fly" extracting infor-mation of the flowing streams, leaving the actual decision-making to the routers. In a second step, COAST provides for content adaptation and optimal distribution over the Future Media Internet.

COAST aims to build a Future Content-Centric Network (FCN) overlay architecture (Fig. 1.8), where the users will just specify which content or service they need, and the COAST framework will find the desired or the most relevant data and forward it to the users in an efficient, timely and network-friendly way.

To realize the above vision, COAST focus on three key innovation pillars:

- "On the fly" identification and distributed "on-line" discovery. COAST creates a content-aware network of intelligent nodes (edge routers, home gateways, terminal devices), which will (a) "on the fly" identify/classify content and identify Web services via inspection of the traffic that flows through them and (b) discover "on line", where services are located and content is located/cached, in order to optimal match users' requests with availability, while meeting specific Service Level Agreements (SLAs) in content consumption;
- Content-Aware Delivery Network Architecture. COAST efficiently and dynamically discovers the underlying network infrastructure as well as user terminal devices and user needs considering the (possibly variable) capacity and quality (especially variable with mobility), the actual foreground (multimedia distribution) and background (other traffic) load, and construct content-aware overlays to offer distributed, robust and network-/service-provider friendly

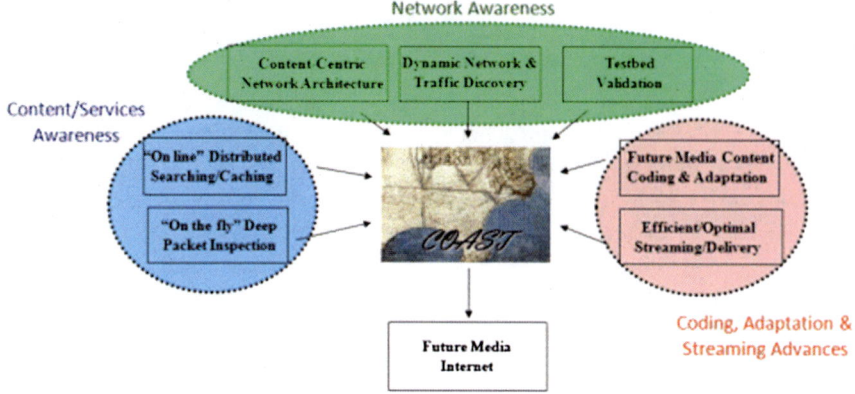

Fig. 1.8 Proposed overlay architecture of (COAST 2012)

content delivery with optimal utilization of the networking topology and resources;
- Future media content adaptation and enrichment. COAST provides for scalable, high-definition 3D/free-viewpoint video with interactive virtual panning/zooming, which will be on-the-fly adapted, enriched and optimized to the user preferences, network and terminal characteristics and conditions and optimal streamed over dynamic constructed overlays.

COAST provides a number of service scenarios, which tries to demonstrate the COAST added value and extract the major COAST functionality and core COAST services. The scenarios created are the following:

- Home based scenarios—this group of service scenarios mainly correspond to Internet access and infotainment consumed or generated from the home environment;
- On-the-move scenarios—this group of scenarios describe the COAST functionality that target mobile devices such as laptops, PDAs or mobile smart phones (with calendar and sensory support);
- Social Networks/Enterprise Related Scenarios – this group of scenarios include the utilization of COAST in enterprise and social environments.

1.4.5 SAIL

SAIL's objective is the research and development of novel networking technologies using proof-of-concept prototypes to lead the way from current networks to the Network of the Future. SAIL leverages state of the art architectures and technologies, extends them as needed, and integrates them using experimentally-driven research, producing interoperable prototypes to demonstrate utility for a set of concrete use-cases. According to the SAIL consortium (SAIL 2012) the SAIL's architecture reduces costs for setting up, running, and combining networks, applications and services, increasing the efficiency of deployed resources (e.g., personnel, equipment and energy).

SAIL aims at improving application support via an information-centric paradigm, that replaces the old host-centric one, whilst developing for that purpose concrete mechanisms and protocols to realize the benefits of a Network of Information (NetInf). SAIL enables the co-existence of legacy and new networks via virtualization of resources and self-management, fully integrating networking with cloud computing to produce Cloud Networking (Murray et al. 2012). SAIL embraces heterogeneous media from fiber backbones to wireless access networks, developing new signaling and control interfaces, able to control multiple technologies across multiple aggregation stages, implementing Open Connectivity Services (OConS).

In SAIL the interaction patterns of emerging applications no longer involve simply exchanging data end-to-end. These new patterns are centered on pieces of information, being accessed in a variety of ways. Instead of accessing and manipulating information only via an indirection of servers hosting them, putting information objects at the centre of networking is appealing and has resulted in the notion of a Network of Information. A Network of Information simplifies communication compared to end-to-end or store-and-forward networking, making it more suitable for large-scale, complex applications and for improved programmer and network efficiency. By caching multiple, equal copies of the same information object, a Network of Information natively supports large-scale content distribution. These properties also make a Network of Information suitable for delay tolerant networking. The Network of Information architecture natively supports mobility of information objects, nodes, users, and applications. But it is still unclear how to make sure that such a Network of Information-based communication model can be supported efficiently in the network at runtime. For example, the model mandates that information be available near to its point of consumption, but that entails caching inside the network—it is not clear how to integrate such caches in routers, especially as the overall network scales up. However, such integration will provide substantial benefits, e.g., by reducing information access delay and required bandwidth, improving dependability, and simplify network operation.

Furthermore, network applications can fluctuate rapidly in popularity and in terms of the amount of user interaction. This makes provisioning a difficult problem, both on the server and storage side, as well as on the networking one. On the server and storage side, cloud computing has successfully addressed many of these challenges, using virtualization as a core technique. However, it is still unclear how to provide suitable network support for such highly variable applications when they run not just over the tightly-controlled, custom-tailored network of a cloud computing operator, but inside more complex, more diverse telecommunication operator networks. Even if it is possible to provide the computational resources, it is not obvious how to dynamically provide the necessary networking support/capacity or the complex networking topology required by, e.g., a multi-tier web application. Nor is it obvious how to provide networking support for distributed cloud systems, spanning multiple networks and operated by different entities. A complex example scenario would be a Network of Information not only serving static, but dynamic information (e.g., results of database queries made to travel websites, or cookie-like state of a distributed application), requiring computing resources dynamically allocated and networked in a flash network slice when the Network of Information entities are reaching capacity limits.

In principle, the raw technologies for efficiently transporting even very large amounts of data are already in place (e.g., optical circuit switching), but these do not mesh well with current Internet control concepts (e.g., packet switching and connection-less networking). Indeed, current solutions struggle with the heterogeneity of the deployed networks and cannot seamlessly integrate control of these resources from the edge (i.e. using the context known to the application). For example, it is practically impossible to exploit the diversity of communication technologies used at two endpoints by switching between technologies, as the rate of the data flow changes (e.g., random variations in channel quality or structural differences in channel properties like different delay/data-rate tradeoffs). Similarly, efficient multi-path/multi-protocol/multi-layer optimization is still infeasible. Cloud Networking, and especially Network of Information, will substantially change the traffic patterns and data flows in the Internet and make the requirement for agile connectivity service even higher.

SAIL deals with particular aspects of a real, complex network. Figure 1.9 shows how entities are all reflected differently in Network of Information, Cloud Networking, and Open Connectivity Services, yet all three aspects stay connected and in relation to each other. The figure also highlights that one expects the Network of Information and Cloud Networking aspects to appear multiple times, while the Open Connectivity Services one constitutes a single basis for the other two.

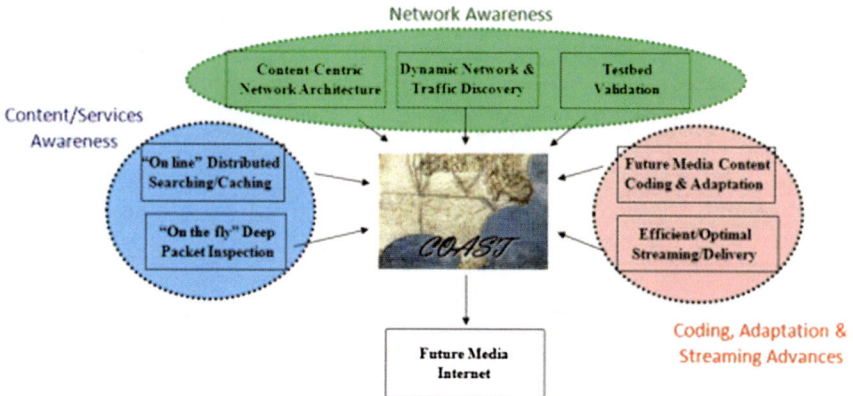

Fig. 1.9 SAIL's three main technical objectives (SAIL 2012)

1.4.6 COMET

The COntent Mediator architecture for content-aware nETworks (COMET) intends to take a unified approach to content location, access and distribution, irrespectively of the intermediary used. COMET introduces a global naming scheme and optimizes both content source selection and distribution, by mapping the content to the appropriate network resources based on transmission requirements, user preferences, and network state. Content and network-aware mapping will consider current, evolutionary and revolutionary network approaches (COMET 2012). The architecture addresses two key aspects:

- Global content naming and addressing, with supporting infrastructure for search and content name resolution to the relevant identifiers required for access;
- Unified access and user-, content- and network-aware distribution by mapping the content onto the appropriate network resources through supporting infrastructure, providing content- and network-aware access to every type of content through all possible types of distribution media.

There are two major conceptual entities in COMET: Content Mediation Server (CMS) and Content Aware Forwarder (CAF).

The CMS is a key conceptual entity introduced by COMET that is responsible for the diverse forms of content manipulation, as for instance, content publication, resolution or delivery operations. All these processes need collaborations between CMS entities, in which case communication protocols are necessary between CMSs for fulfilling specific tasks. Although conceptually the CMS is one entity, in real implementations, it can be a combination of several physical machines with each of them providing separate services but interconnected to form a coherent mediation plane.

The CAF is another key entity for enabling content distribution across the global Internet. CMSs are not necessarily responsible for actually carrying the content from the content source to the individual Content Consumers. Instead, such role is fulfilled essentially by content aware forwarders (CAF entities). Compared to legacy IP routers, content aware forwarders are capable of processing content packets based on their identifiers in a native way, while it is not necessary to force all routers within an ISP's network to be content aware. In such a scenario, content aware forwarders form a virtual network infrastructure for actually delivering the content.

Besides the tow major entities presented above, two other entities have the mission to identify the physical end points involved in the end-to-end content distribution. These are the following:

- Content Server—the server that actually hosts content is referred to as Content Server. It should be noted that Content Server are typically maintained by Content Providers, and hence it does not belong to the COMET system;

- Content Client—the actual end host machine, which is the destination of a content flow. In practice, Content Clients are owned by Content Consumers. Like the entity above, also Content Client is not an integral part of the COMET system.

COMET system offers two types of content-based operations: content publication (Content Provider oriented) and content consumption (Content Consumer oriented). Figure 1.10 schematizes these processes and represents the involved entities.

The content publication is the process of making content available to Content Consumers. Replicas of a specific content object are stored in content server owned by Content Providers. Once a Content Provider has new content, it needs to notify the COMET system, specifically the CMS entity, for making it available for access across the Internet. On the other side, content consumption is the process initiated by a Content Consumer to receive the requested content. This process is mostly transparent to the Content Consumer who, through a Web browser, just needs to click on a hyperlink or to type in the corresponding URL.

The COMET project follows a 2-plane approach, including a Content Mediation Plane (CMP) and a Content Forwarding Plane (CFP), as shown in Fig. 1.11.

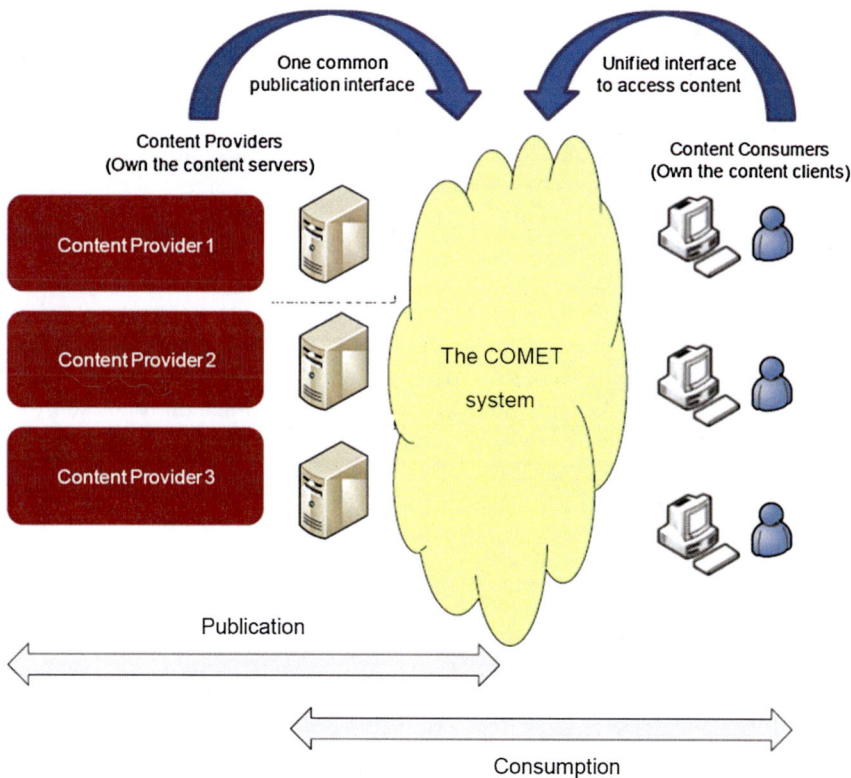

Fig. 1.10 Schema of content publication and content consumption (COMET 2012)

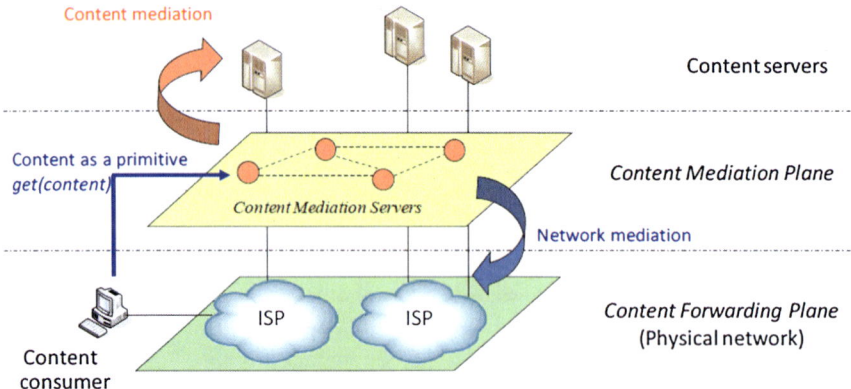

Fig. 1.11 COMET two-plane approach (COMET 2012)

The basic task of CMP is to locate the actual content server that hosts the requested content. In case multiple content servers have the requested content, it is the responsibility of the CMP to find the best candidate according to specific criteria, such as server load conditions and/or the associated delivery path quality. From this point of view, both the aforementioned name resolution and content resolution functions are typically enforced in the CMP.

The Content Forwarding Plane (CFP) is in charge of delivering the content, once requests are resolved based on its current knowledge of both the network and server status. Thus, once the requested content has been found (or the best copy is identified, when there are available multiple copies of the same content), the CFP will be responsible for enforcing the actual path back to the Content Consumer from the identified content source.

The two planes (CMP and CFP) are not independent of each other and are required to communicate to achieve smooth and complete operation of content access (i.e. from the request issued by the Content Consumer to the final delivery of the content back to the requester). As an example for such a vertical interaction between the CMP and the CFP, the CMP will require information from the CFP to achieve network-awareness, whereas, on the reverse direction, the CFP will require information from the CMP in establishing the most suitable paths for the transportation of content with specific QoS requirements. This bi-directional interaction of both planes is instrumental in achieving a seamless content delivery.

1.4.7 CONVERGENCE

CONVERGENCE advocates that the original Internet conceived as a "network of hosts" is evolving into an Internet of services, an Internet of media, an Internet of people and an Internet of things. This implies a strategic shift from "host-centric" to "content-centric" and "data-centric" networking. Against this background, CONVERGENCE proposes to enhance the Internet with a novel, content-centric, publish-subscribe service model, based on the concept of Versatile Digital Item (VDI): a common container for all kind of digital content, derived from the MPEG21 standard (CONVERGENCE 2012).

The VDI is a "package" of digital information with a unique, host-independent, self-certifying identifier, capable of encapsulating any kind of digital information, including information on media, services, people and "real world objects". A VDI also contains metadata, describing its content. VDIs can be used both to publish and to subscribe to content. Networking functionality allows storage and routing of VDIs through the network. An open source middleware exposes network functionality to information providers and consumers, allowing them to search for, retrieve, modify and revoke VDIs. CONVERGENCE makes it possible to maintain access to VDIs when they move from one host to another, protect user security and privacy, and provide support for "digital forgetting", ensuring that VDIs are deleted when they pass a user-defined expiry date. An RDF-based Community Dictionary Service supports search for user-defined metadata. CONVERGENCE concepts have been tested in iterative cycles of research, design and experiment, studying their usefulness in real-life business scenarios, integrated into partners' product cycle, and using large-scale FIRE facilities to investigate their scalability and robustness.

CONVERGENCE targets professional and non-commercial providers and consumers of digital content, allowing them to publish, control, search for, and use content, independently of the structure or geographical location of the content. Users will be able to define their own policies for using, authenticating, protecting and revoking VDIs. The functionality provided by VDIs supports new models of use and new business models, difficult or impossible to implement on the current Internet architecture.

The next chapters of this book will analyze with some level of detail the architecture of CONVERGENCE, its main functionalities and implementation strategy.

1.5 Conclusions

Information Centric Networking is one of the major themes driving research challenges in the Future Media Networks over the next decade. As applications become more distributed and interactive, many of the traditional network functions of naming, addressing, routing and forwarding become strongly influenced by the applications and the related content/media.

Currently, the originally "best effort" principle of Internet operation is not good enough to deliver the QoE end-users want. Furthermore, streaming audio and video, file downloads, online games, software-as-a-service and social networking demand different quality parameters such as bandwidth and network delay, and such selectiveness is not natively supported in the current Internet.

CDNs need to be customized according to the content they deliver, which implies that there is no single solution for all applications. In addition, CDNs are typically used by professional content creators and are not for free, which leaves small players and non-professional providers without a content distribution service.

The future approaches should try to harmonize the overall CDNs, rather than focusing on individual network entities, in order to support an end-to-end QoS and QoE architecture over heterogeneous networks, applied to a variety of audio-visual services that are delivered to various user terminals. Furthermore, future approaches shall mitigate the current main challenges of CDN architecture in terms of security, efficient algorithms for load balancing, content caching and replication.

References

4WARD, "4WARD – Architecture and Design for the Future Internet". Available: http://www.4ward-project.eu/index.php [Accessed: October 11, 2012].

A. Carzaniga, M. J. Rutherford, A. L. Wolf, "A Routing Scheme for Content-Based Networking". In Proceedings of the The 23rd Conference of the IEEE Communications Society, Hong Kong, March 2004. Available online: http://www.inf.usi.ch/carzaniga/papers/crw_infocom04.pdf [Accessed: March 2013].

A. Detti, M. Pomposini, N. Blefari Melazzi, S. Salsano and A. Bragagnini, "Offloading cellular networks with Information-Centric Networking: the case of vídeo streaming", IEEE International Symposium on a World of Wireless, Mobile and Multimedia Networks 2012 (WoWMoM 2012).

A. Feldmann, "Internet clean-slate design: what and why?", In Proceedings of SIGCOMM Computing Communications, pp. 59–64, 2007.

A. Mitra, M. Maheswaran, "Wide-Area Content-based Routing Mechanism". In Proceedings of the International Parallel and Distributed Processing Symposium (IPDPS'03), Nice, France, April 2003.

A. Pathan, "A Taxonomy and Survey of Content Delivery Networks", Grid Computing and Distributed Systems Laboratory, University of Melbourne, Australia, 2007.

B. Ahlgren, C. Dannewitz, C. Imbrenda, D. Kutscher and B. Ohlman, "A Survey of Information-Centric Networking", IEEE Communications Magazine, IEEE Com. Mag., Vol. 50 n°7, 2012, Special issue on ICN, July 2012.

COAST, "COAST – Content Aware Searching retrieval and sTreaming". Available: http://www.coast-fp7.eu/ [Accessed: October 11, 2012].

COMET, "COMET—COntent Mediator architecture for content-aware nETworks", Available: http://www.comet-project.org/index.html [Accessed: October 17, 2012].

CONVERGENCE, "The CONVERGENCE project". Available: http://www.ict-convergence.eu/ [Accessed: October 23, 2012].

D. Perino and M. Varvello, "A reality check for content centric networking", In Proceedings of ACM SIGCOMM Workshop, pp. 44–49, 2011.

D. Sjöberg, "Content Delivery Networks: Ensuring quality of experience in streaming media applications", TeliaSonera International Carrier, 2008.

D. Smetters and V. Jacobson, "Securing network content", PARC TR White Papers, pp. 1–7, 2009.

F. Salguero, "Content Mediator Architecture for Content-aware Networks", COMET EU FP7 Report, 2010.

FIA – Future Content Networks Group, "Why do we need a content-centric future Internet?", EC, Information Society and Media, 2009.

G. Carofiglio, M. Gallo, L. Muscariello and D. Perino, "Modeling data transfer in contente-centric networking", In Proceedings of 23rd International Teletraffic Congress, pp. 12–23, 2011.

G. Pallis, "Insight and Perspectives for Content Delivery Networks", Communications of the ACM, 49(1), pp. 101–106, 2006.

G. Polyzos, "Context-Aware Information Delivery in Assistive Environments over a Publish-Subscribe Internet", In Proceedings of the 1st International Conference on Wireless, pp. 67–71, 2011.

G. Tyson, "A Middleware Approach to Building Content-Centric Applications", PhD thesis, Lancaster University, 2010.

H. Moustafa and S. Zeadally, Media Networks: architectures, applications and standards, CRC Press, 2012.

IETF, Internet Engineering Task Force RFC 6770, "Use Cases for Content Delivery Network Interconnection", November 2012.

J. Broberg, R. Buyya, and Z. Tari, "Metacdn: Harnessing storage clouds for high performance content delivery", Networked Computer Applications, 32(5), pp. 1012–1022, 2009.

J. Höller and J. Arkko, "Internet of things propels the networked society", June 14, 2012. [Online]. Available: http://labs.ericsson.com/ [Accessed: September 12, 2012].

M. Blumenthal and D. Clark, "Rethinking the design of the Internet: the End-to-End arguments vs. The Brave New World", ACM TOIT, pp. 70–109, 2001.

M. Demmer, B. Du, A. Ermolinsky, K. Fall, I. Ganichev, T. Loponen, J. Liu, S. Shenker and A. Tavakoli, "A publish/subscribe communications api for data-oriented applications", In Proceedings of the 5th Symposium on Networked Systems Design and Implementation, pp. 94–102, 2009.

M. Grafl, C. Timmerer, H. Hellwagner, G. Xilouris, G. Gardikis, D. Renzi, S. Battista, E. Borcoci, D. Negru, "Scalable Media Coding enabling Content-Aware Networking". In IEEE Multimedia, 27 Nov. 2012. IEEE computer Society Digital Library. IEEE Computer Society, (http://doi.ieeecomputersociety.org/10.1109/MMUL.2012.57).

N. Blefari Melazzi, L. Chiariglione, "The potential of Information Centric Networking in two illustrative use scenarios: mobile video delivery and network management in disaster situations", invited paper, IEEE Journal MMTC, E-letter special issue on "Multimedia Services in Information Centric Networks".

P. Carvalho, M. Andrade, C. Alberti, H. Castro, C. Calistru and P. Cuetos, "A unified data model and system support for the context-aware access to multimedia content", In Proceedings of the Workshop on Multimedia Semantics, pp. 72–93, 2007.

P. Murray, A. Sefidcon, R. Steinert, V. Fusenig, J. Carapinha, "Cloud Networking: An Infrastructure Service Architecture for the Wide Area", In Proceedings of the Future networks and Mobile Summit (FuNeMS 2012), Berlin, Germany, July 2012.

PURSUIT, "PURSUIT – Publish Subscribe Internet Technology". Available: http://www.fp7-pursuit.eu/ [Accessed: October 4, 2012].

R. Buyya, M. Pathan, A. Vakali (Eds.), "Content Delivery Neyworks". Springer Series Lecture Notes in Electrical Engineering, Vol. 9, 2008, ISBN 978-3-540-77887-5.

R. Jian, "Internet 3.0: Ten problems with current Internet architecture and solutions for the next generation", In Proceedings of MILCOM 2006, pp. 1–9, 2006.

SAIL, "SAIL – Scalable and Adaptive Internet Solutions". Available: http://www.sail-project.eu/ [Accessed: October 15, 2012].

SKA, "Over-The-Top Content Distribution To Grow Substantially". Available: http://seekingalpha.com/article/1039521-over-the-top-content-distribution-to-grow-substantially [Accessed: December 10, 2012].

S. Oueslati, J. Roberts and N. Sbihi, "Ideas on Traffic Management in CCN", Information Centric Networking, Dagstuhl Seminar, 2011.

T. Koponen and M. Chawla, B. Chun, A. Ermolinskiy, K. Kim, S. Shenker and I. Stoica, "A data-oriented (and beyond) network architecture, Proceedings of the SIGCOMM, Kyoto, Japan, pp. 1–12, 2007.

T. Roscoe, "The end of the Internet architecture", HotNets White Papers, pp. 1–6, 2006.

V. Jacobson, K. Smetters, J. D. Thornton, M. F. Plass, N. H. Briggs, R. L. Braynard, "Networking named content", In Proceedings of ACM CoNEXT, pp. 1–12. 2009.

Y. Li, Z. Li, M. Chiang and A. R. Calderbank., "Content-Aware Distortion-Fair Video Streaming in Congested Networks". In IEEE Transactions on Multimedia, Vol. 11, No. 6, October 2009.

Chapter 2
CONVERGENCE Architecture a Concise Overview

Maria Teresa Andrade, Nicola Blefari Melazzi,
Charalampos Patrikakis and Richard Walker

Abstract Conceived as a network of Hosts, the Internet is evolving into a network of content and services. To meet new demands associated with this evolution, CONVERGENCE proposes a novel content-centric architecture, supporting a publish-subscribe paradigm for this enhanced Internet. The strategic goal of the project is to design, test and refine this framework using one or more available experimental facilities. This chapter starts by providing a description of the major concepts and challenges that guided the conception and design of the CONVERGENCE architecture. It then describes, with some level of detail, the devised solution, in terms of its components, corresponding high-level functionality and hierarchy. It concludes by illustrating how such architectural components can be used, with the presentation of concrete deployment scenarios.

2.1 Introduction

The Internet was developed with a host-centric approach. However, it is now evolving into a direction where the key elements will no longer be "hosts", but rather exchanged data and services (Aitenbichler et al. 2010, Zahariadis et al. 2010). In this process, the amount and variety of resources available on-line, which has already significantly increased, is expected to continue to rise, turning search

M. T. Andrade (✉)
INESC TEC, Faculty of Engineering, University of Porto, Porto, Portugal
e-mail: mandrade@fe.up.pt

N. Blefari Melazzi
Electronic Engineering Department, University of Rome "Tor Vergata", Rome, Italy

C. Patrikakis
School of Electrical and Computer Engineering, NTUA, Athens, Greece

R. Walker
Blue Brain Project, Ecole Polytechnique Federale de Lausanne, Lausanne, Switzerland

F. Almeida et al. (eds.), *Enhancing the Internet with the CONVERGENCE System*,
Signals and Communication Technology, DOI: 10.1007/978-1-4471-5373-3_2,
© Springer-Verlag London 2014

and retrieval of useful content into an extremely difficult task for the average Internet user. Within this context, it is of utmost importance to allow the efficient establishment of relationships between comparable resources and to allow search engines to use those relationships with the aim of helping users to find resources of their interest.

The current Internet architecture and the services that are made available through it, are not fully adequate to address these new challenges in an optimal manner, nor are they prepared to allow on-line businesses to be deployed taking full advantage of Internet's overall potential. In this context, the CONVERGENCE project proposes to develop key enhancements to the current Internet architecture, to support and foster the referred on-going Internet evolution. Specifically, CONVERGENCE introduces:

- the concept of the Versatile Digital Item (VDI), an extension of the MPEG-21 Digital Item (DI) (MPEG-21 2005), acting as the fundamental unit of transaction and distribution of resources, representing information about media, Real World Objects, services and people;
- a new networking functionality supporting publication/subscription, discovery, routing and authentication of VDIs;
- a new middleware that operates on top of the networking mechanisms, supporting and taking advantage of its novel functionality whilst pervasively employing the VDI container. This middleware platform offers application developers an open set of tools to deploy their own applications over it; the middleware includes a Community Dictionary Service, supporting semantically accurate search;
- novel security and privacy mechanisms including support for "digital forgetting".
- applications, relevant to the needs of business and educational organizations participating in the project, and tools, which are re-usable elements, facilitating re-use of code in applications: an application can make use of several tools.

Based on the identified challenges and on the substantiation of the paradigm shift currently being observed in the Internet, CONVERGENCE has set as a major goal, the enhancement of the Internet by means of a content-centric, publish-subscribe service model Eugster et al. (2003). Simultaneously, and contributing to this aim, it defined and built the necessary functionality to manipulate a novel format for the declaration and representation of resources on the Internet.

This new format, the Versatile Digital Item (VDI), is able to support all information objects and services, as well as representation of people and Real World Objects (RWOs). VDIs are the basic packaging, distribution and transaction unit in the CONVERGENCE system and hence in the new Internet architecture that CONVERGENCE advocates. They can incorporate any kind of information, including, for instance, signaling and control information. This minimizes the need for information storage outside this data unit (though this is still possible).

The introduction of VDIs is one of the key features of the new Internet solution envisioned by CONVERGENCE. This choice is coherent with the current trend of "content-centric" networking, whereby the network layer provides users with content, instead of communication channels between hosts, and is aware of this content, i.e. it knows at least its name. This shift is analogous to the shift from circuit to packet switching. In circuit switching a PCM slot contains only user data whereas in packet switching, an IP datagram also contains (among other things), a destination address. Similarly, in "content switching", the VDI contains a complete package of user data and meta-data describing content and how to handle it.

As previously explained, the VDI container can be used to declare and package any kind of digital resources, such as media files or data about services. It is also able to perform the declaration of people and RWOs, such as, for instance, items of merchandise identified with an RFID. Furthermore, VDIs bind meta-information to the declared resources, in order to describe the context, content and structure of such resources (other VDIs, audio, images, video, text, descriptors of RWOs, descriptors of people etc.). The employed meta-data include: structural information for the description of the composition of VDIs; semantic information to provide a characterization of the declared resources; cryptographic information to enable the robust authentication and protection of information included in the VDIs; usage rights information for the definition of users' rights to manipulate the VDIs and their resources; expiration information to provide support for "digital forgetting"; and identification information to enable the universal differentiation of all VDIs as well as their logical binding to their corresponding network-level information packets.

Another key feature of CONVERGENCE is the support of a publish/subscribe service model. Within such a model, subscribers register their interest for a subject, or for a pattern of subjects, and are asynchronously notified of relevant resource publications, (done by publishers). The publish/subscribe service model effectively decouples the application end points in both space and time. This decoupling enables a greater systemic scalability, a more dynamic network topology and a much enlarged and flexible typology of services.

Figure 2.1 depicts the main players in the CONVERGENCE pub/sub framework, i.e. publishers and subscribers. Publishers advertise their resources (data and service-access-points) on the CONVERGENCE system, by way of publications. Subscribers register, within the system, their interest in specific resources or topics. The system notifies subscribers when resources matching their interests become available. Users can also search for resources, in which case the system immediately provides a response. A search can be seen as a special form of subscription where the response is synchronous to the request.

The publication and subscription procedures in CONVERGENCE naturally employ VDIs. In a publication event, a resource is published by creating its VDI and injecting it onto the CONVERGENCE system. Publication VDIs contain meta-information as listed above, among which information that semantically characterizes the declared resources. Subscriptions are performed by creating a VDI containing the criteria that represent the subscriber's needs and preferences. Subscription VDIs thus contain meta-information that identifies the users' needs and preferences. The CONVERGENCE system compares subscription VDIs with

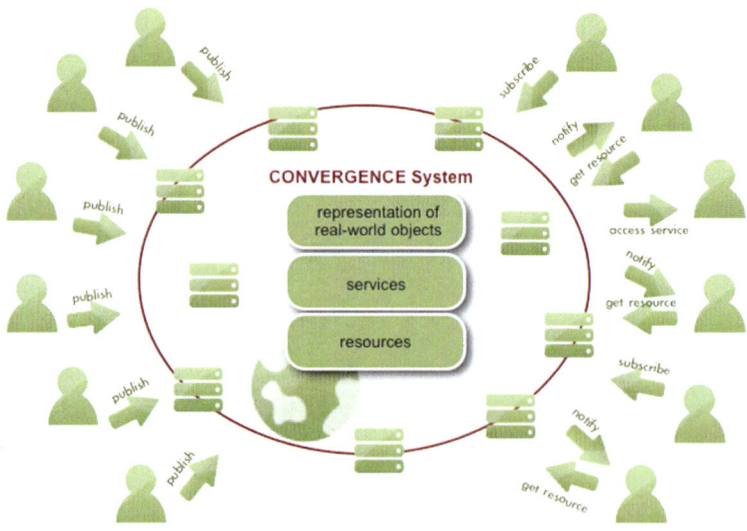

Fig. 2.1 Service model

publication VDIs to detect matches and subsequently notify the relevant subscribers of the existence of resources meeting their requirements.

The VDI object is thus seen by CONVERGENCE as the vehicle through which the future Internet will provide the support for a whole set of operations inherent to the publication, subscription, access to and transaction of any kind of resources within a publish-subscribe operational paradigm. The CONVERGENCE architecture delivers a comprehensive, distributed and extensible set of open software engines, that collectively provides the necessary functionality to manipulate VDIs, enforcing a publish-subscribe operation and enabling the implementation of a broad range of services on top of it.

2.2 The Layered Architecture

CONVERGENCE specifies an Information and Communication Technology (ICT) system adopting a three-level architectural model. It comprises a set of innovative technologies, partially based on open standards, which cooperate to offer the content-centric publish-subscribe functionality referred to in the previous section.

Figure 2.2 illustrates the layered approach adopted by CONVERGENCE for the new Internet architecture that it advocates. The picture depicts the three functional levels of the CONVERGENCE system (Application, Middleware and Computing Platform), summarizing the scope of each level and the kind of information exchanged at the interfaces between levels. Each such functional level is an aggregate set of conceptually similar functionality.

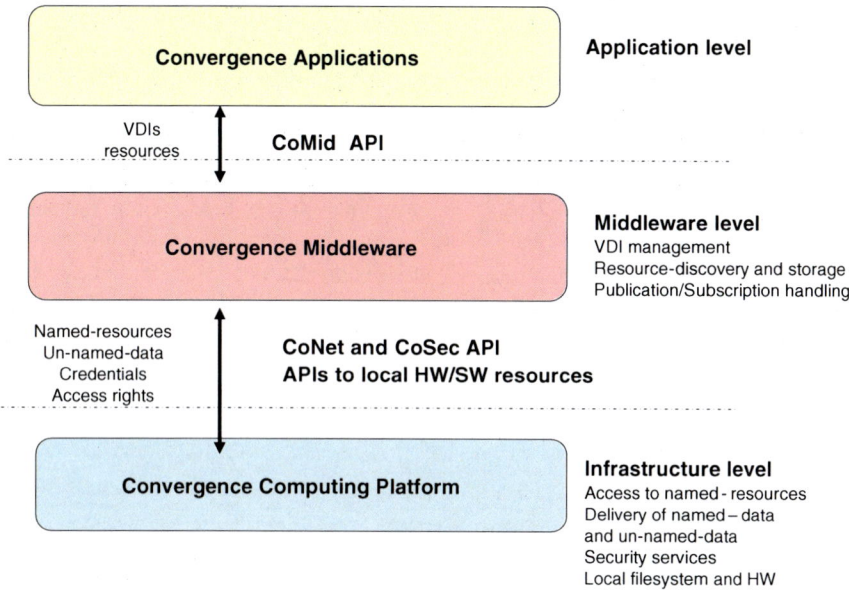

Fig. 2.2 Overview of CONVERGENCE architectural levels

2.2.1 Application Level

The set of all CONVERGENCE applications comprises the Application Level, or CoApp. These applications, which can be freely downloaded and installed, provide any user with the means to create, process, manage and consume VDIs and their components and resources, albeit not requiring that users have specific knowledge on VDIs. They are the direct vehicle through which users can flexibly manipulate digital resources and deal with physical objects and their digital counterparts, as well as with people and their digital identities.

The CoApp level thus provides the mechanisms to transparently exploit the functionality offered by the services running at the lower levels of the CONVERGENCE architecture, specifically, at the Middleware level. Creators who want to make their resources available on the CONVERGENCE network need to generate corresponding publication VDIs. This is accomplished via CoApp mechanisms. Once created, these VDIs are injected in the network through the Middleware level, where they undergo further manipulation. This step allows the system to extract meta-data, which is crucial for enabling VDIs to be subsequently searched and accessed. Accordingly, the CoApp and Middleware levels exchange resources and VDIs at the interface between them. For that purpose, the Middleware level offers a standard Application Programming Interface (API) known as the CoMid API. Employing it, applications can resort to the services of local or remote middleware instances.

To expedite the development of applications, CONVERGENCE defines special re-usable application elements called tools, which facilitate the re-usage of code across applications. The CoApp level is thus logically divided in two sub-levels: a User Applications sub-level and a Tools sub-level. Therefore the term Application can be used in a wider sense to refer to both User Applications and Tools.

As illustrated in Fig. 2.3, five major Application functionalities and corresponding Tools have been defined within the CoApp level. They have been identified as the ones that provide the core functionality needed by the vast majority of high-level applications populating the future Internet. Further functionality may be provided by accessing the services offered by the middleware or through the development of new Tools.

2.2.2 Middleware Level

The CONVERGENCE Middleware (CoMid) level enable and handle the creation, retrieval, manipulation and consumption of VDIs. The CoMid allows users who create resources (User-producers) to submit and publish their creations as VDIs, making them available to all other users (User-consumers). Upon publication of such VDIs, CoMid provides CONVERGENCE users with the means to search for them according to semantically precise search criteria. CoMid also handles the

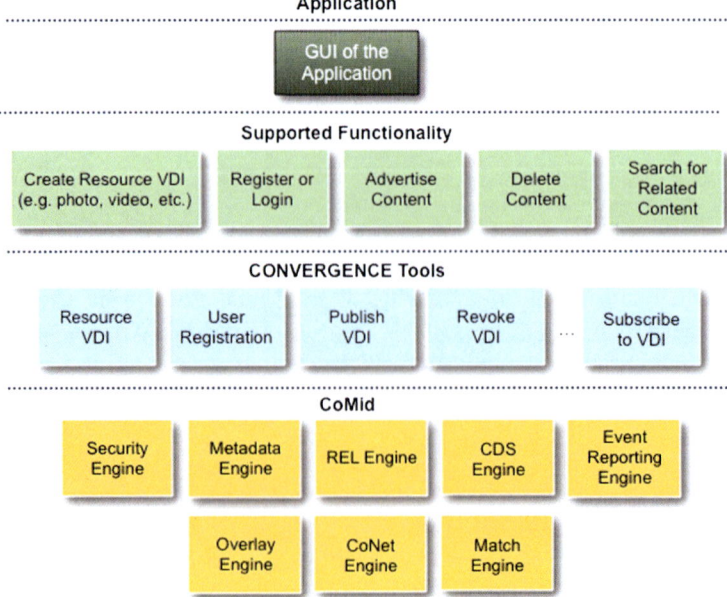

Fig. 2.3 The CoApp level and its interface to the CoMid

registering of User-consumer subscriptions, their matching to incoming publications and the automatic notification of relevant User-consumers when VDIs
matching their interests become available. The CoMid is thus responsible for the
delivery of search results and notifications to the Application level.

CoMid is based on the MPEG-M standard (MPEG-M 2013 a, b, c). This
standard provides a distributed eco-system of Engines, i.e. components that can be
activated by Applications running on MPEG-M compliant devices. Similarly,
CoMid defines a set of Engines providing services that are exposed to Applications
running on CONVERGENCE devices. The set of Engines defined by CoMid
builds on and expands the MPEG-M design. Specifically, new Engines are
introduced by CONVERGENCE to deliver the following functionality, non-
existent in the MPEG-M standard:

- implementation of the publish/subscribe paradigm on top of MPEG-M Event
  Notification services;
- possibility to perform semantically precise search for and subscription to published resources.

There are two main classes of Engines:

- Protocol Engines (PE) that activate functionalities in remote or local peers;
- Technology Engines (TE) that are typically called by PEs to execute specific
  operations.

PEs can be seen as the coordinators of specific procedures to be performed over
VDIs. These engines configure, interconnect, trigger and use TEs to perform such
procedures. TEs can be seen as the unitary operational components that make up
the PEs. They implement specific unitary operations and cooperate amongst
themselves to deliver the complete desired functionality of the PE.

PEs are associated to Elementary Services (ESs). An ES, a concept imported
from the MPEG-M standard (MPEG-M 2013tris), consists of a basic unit of
functionality, offered by MPEG-M, having a specific value or meaning to the
Application or User. An ES constitutes a pillar functionality of MPEG-M and,
likewise, of CoMid.

In functional terms, the CoMid is composed of two major building blocks: the
block providing functionality to manipulate VDIs and the block providing functionality to enable semantically-rich processing of VDIs.

The former block is composed by a number of MPEG-M defined Engines, as
well as by a limited number of CONVERGENCE-specific Engines. Among those
that have been adopted from MPEG-M, some are used in their original format,
whereas others have been extended to adapt to requirements identified in
CONVERGENCE.

The latter block comprises essentially two components: the CONVERGENCE
Core Ontology (COO) and the Community Dictionary Services (CDS). The CCO
enables the creation of a semantic overlay, organized into fractal spaces with
precise semantic meaning, for an efficient distribution of VDIs. The CDS is

implemented as a TE and it enables the translation between ontologies (more specifically, between external ontological models and the CCO's and vice versa). It thus enables the CoMid to support ontologies with concepts and properties different from those defined in the CCO, and, consequently, enables users to develop and employ dedicated ontologies when describing their resources.

Being an intermediate level, CoMid offers two APIs. Through them, applications at the upper level can invoke functionality offered by CoMid, as explained in the previous sub-section. At the lower level, CoMid may request services from the Computing Platform level. VDIs are exchanged through these interfaces in both senses: from top to bottom, upon creation of resources and subsequent encapsulation into VDIs, as well as when publishing and distributing these VDIs within the network; from bottom to top, when accessing, retrieving and consuming VDIs previously published into, and disseminated throughout, the network.

An application call may involve more than one PE or TE. Therefore CONVERGENCE provides a standard mechanism for "aggregating" PEs and "orchestrating" TEs. A similar approach can also be adopted to manipulate applications.

Figure 2.4 depicts how an application call may involve a chain of PEs and TEs. Applications normally access the services of CoMid by invoking specific PEs. These may require the services of other PEs and TEs forming cooperating engine chains, where the participating elements communicate via the CoMid API. However, chains need not be linear: where useful, applications can directly call TEs without intermediation by a PE.

At the lower level, CoMid provides direct interfaces to the two key components of the Computing Platform: CoNet and CoSec. To that aim it includes two specific TEs, the CoNet TE and the Security TE, implementing counterpart functionality at the middleware level.

2.2.3 Computing Platform Level

The CONVERGENCE Computing Platform Level (CoComp) is responsible for the provision of networking and security functionalities, as well as for performing the interfacing to local resources, (e.g. file-systems, processing power, etc.). It comprises two key functional blocks providing network level services, namely, novel content-centric networking (CoNet) and secure handling of resources within the networked environment (CoSec). The Computing Platform Level handles the

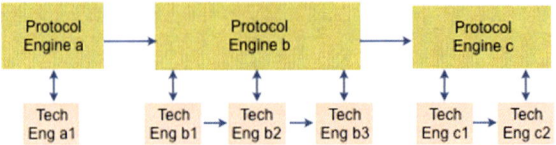

Fig. 2.4 A chain of protocol and technology engines

storage, routing and retrieval of resources in a reliable manner. Also, it provides security features to simultaneously enable the controlled access to resources, the protection of their integrity and of their privacy. The former set of features is provided by the CoNet component, whereas the latter is mostly provided by the CoSec component.

CoNet is an inter-networking technology that interconnects underlying CoNet Sub Systems (CSS), in the same way as IP interconnects the sub-networks that collectively form the current Internet (Detti et al. 2011). A CSS contains CoNet nodes and exploits an under-CoNet technology to transfer data among CoNet nodes. A CSS could be (see Fig. 2.5, for more details see Chap. 3 and Detti et al. 2011):

- a couple of nodes interconnected by a point-to-point link, e.g. a PPP link or a UDP/IP overlay link;
- a layer-2 network, e.g. Ethernet, or a layer-3 network, e.g. a private/public IPv4 or IPv6 network, or a whole IP Autonomous System, or even the whole current Internet.

The devices within a CSS use an autonomous and homogeneous *under*-CoNet addressing space and, if necessary, an interior *under*-CoNet routing protocol. CSSs can be defined rather freely. For instance, if CoNet protocols are implemented only in user equipment, interconnected by the current Internet, then we have only one CSS: the current Internet. If CoNet protocols are implemented in current border gateways (i.e. where BGP runs), then CSSs coincide with current Autonomous Systems. If CONET protocols are implemented in all current routers, then CSSs coincide with current IP subnets. If CoNet protocols are implemented in nodes that interconnect different layer 2 networks, removing IP, then CSSs coincide with such layer 2 networks.

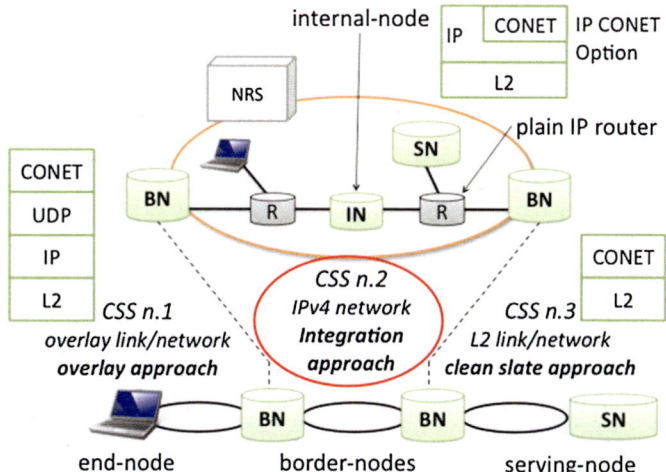

Fig. 2.5 Conet architecture

CoNet nodes are distinguishable according to the functionality they provide when handling data units within CoNet, in regards to:

- forwarding capability;
- caching and verification of cached data;
- data repository;
- address mapping capability;
- reliable data transfer.

CoNet has five different types of nodes, specifically: Border nodes, Internal nodes, Name Routing System (NRS) nodes, Serving nodes and End nodes (see Fig. 2.5). Internal nodes are optional and could be deployed *inside* a CSS to provide in-network caches; differently from border-nodes, internal-nodes forward carrier-packets by using only under-CONET routing mechanisms.

Border, Serving and End nodes all have the capability to forward data units within CoNet. Border and Internal nodes can cache data and verify cached data. Serving nodes provide for permanent secured repositories of data. End nodes assure reliable transfer of data to users. Finally, NRS nodes provide routing information needed to route requests of content to End nodes or Serving Nodes holding that content, on the basis of the Network Identifiers (names of data units) of such desired content.

The overall role of CoNet is to distribute, discover and exchange content objects, divided into named chunks. Thus, to enable the delivery of a specific content, a server node splits it into blocks of data, (the mentioned chunks), assigns a unique Network Identifier (NID) to each such chunk, and disseminates them throughout CoNet. To obtain a chunk, a user issues an Interest message, which contains the name of the desired chunk. CoNet nodes route-by-name the Interest message, using a longest prefix matching forwarding strategy and a name-based routing table. The first en-route device that has the chunk (cached or in a repository), sends it back as part of the response message. CoNet nodes forward it towards the requesting client, through the same sequence of CoNet nodes previously traversed by the Interest message. Downloading a whole item of content is achieved by sending a flow of Interest messages to retrieve all the chunks that compose the content.

CoSec is a technology that implements security services on this networking architecture. The main security features offered by the system, achieved through the cooperation between CoSec, CoNet and CoMid levels, are: the assurance of VDI integrity and authenticity; the governance of VDI access restrictions (confidentiality); the identification and authentication of users; the issuing and enforcing of licenses; the protection of user privacy; and network level security functions. The latter functions regard network level data units (named resources) and tackle several aspects: i) integrity: received content has not been modified, i.e. it is the originally published one; ii) provenance: source of the content is authentic, i.e. the data is provided by the original creator; iii) relevance: received content is really the content requested by the user (for more details see Chap. 3).

CoNet's communication functionality is accessed via the CoNet middleware TE. There are other CoMid engines that use the CoNet TE to perform "raw" CoNet operations (such as Advertise, Get or SendToName). These operations enable a CoMid engine to directly exploit CoNet functionality to handle named-resources. For instance, CoMid engines may use the CoNet TE to interact with CoNet to store and advertise VDIs or any other named resources, so as to make them accessible to other remote middleware entities. CoMid Engines can also employ the CoNet TE to retrieve advertised named-resources. CoNet's functionalities are accessed via the CoNet API.

Credentials and access-rights, extracted from licenses, and embedded within VDIs, are exchanged with the CoSec components of the Computing Platform Level. Applications access these functionalities via the CoMid's Security TE. The Security TE resorts to the services offered, directly, by CoSec via the CoSec API.

2.2.4 Deployment Architecture

A CONVERGENCE system is an implementation of the CONVERGENCE specification, resulting in a set of interconnected peers and nodes collectively called "CONVERGENCE devices".

Peers are devices based on the just described triple-layered functional architecture of CONVERGENCE. From the top down:

- Application level (CoApp)—optional component;
- Middleware level (CoMid)—mandatory component;
- Computing Platform level (CoComp)—mandatory component.

The existence of two mandatory components in peers, implementing the functionality of the two lower levels of the CONVERGENCE architecture, ensures that any peer is able to satisfy the requests of any other peer to manipulate and perform any kind of operation over VDIs.

Nodes are devices that implement only the Computing Platform level, either in its entireness or only partially. Nodes include a CoNet networking component and/ or a CoSec component. There can thus be CoNet and CoSec nodes. The high-level structure of nodes and peers is illustrated in Fig. 2.6.

Figure 2.7 depicts the complete architecture of a peer. Each of the three levels has its own structure and communicates with other levels via standard APIs.

Figure 2.8 provides a distributed view of the CONVERGENCE system, seen as a set of interconnected peers and nodes. The services offered by CoMid Engines, deployed in multiple peers, provide the required functionality to the upper level in a distributed fashion. Network level and security functionality are also realized in a distributed way, employing CoNet and CoSec nodes.

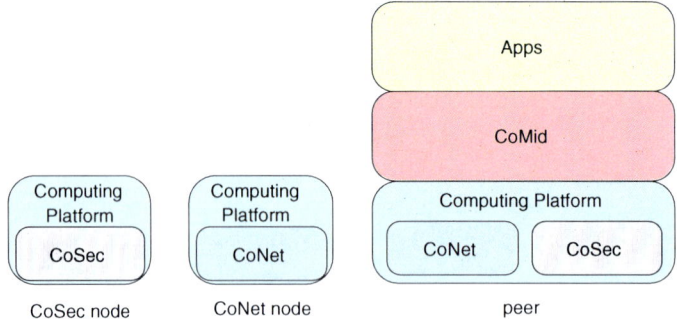

Fig. 2.6 CONVERGENCE nodes and peers

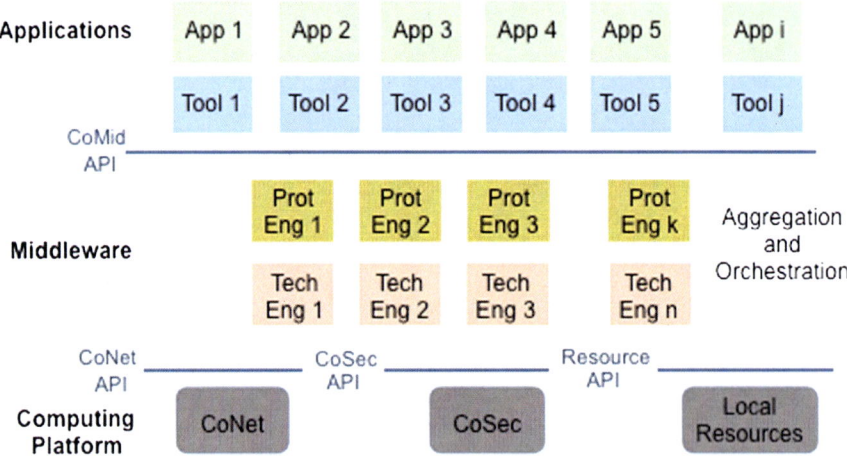

Fig. 2.7 Architecture of a CONVERGENCE peer

2.2.4.1 Peer Deployment Architecture

A CONVERGENCE peer is deployed by installing both the CoMid and CoComp levels in a host. Some peers may also include the CoApp level. This means that the deployment of a peer requires that all CoMid and CoComp software modules be installed and running to offer the functionality associated to those levels.

CoMid is developed in the Java programming language. As such, all peers require a Java Virtual Machine (JVM). Additionally, depending on the engines a peer hosts, it may also be necessary to install the BouncyCastle library for Java security extensions, and the Enterprise Java Beans (EJB) framework. This setup contemporarily provides the support for the Tomcat or JBoss application servers, which are required to host elementary services and applications, as well as smart

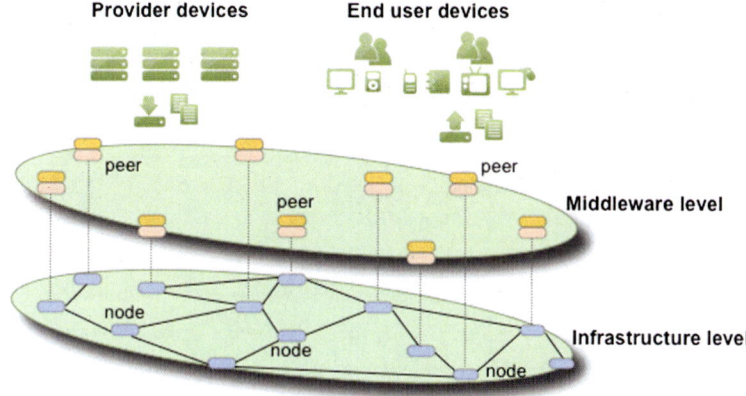

Fig. 2.8 Distributed view of the convergence system

card hardware and software components for the security module. In addition to EJB3 (Enterprise JavaBean3), further libraries/frameworks may be necessary, due to specific requirements imposed by certain engines. These are specified in the engines' own dependencies.

Peers can be deployed on hosts running different operating systems. At present, tests have been conducted on Windows Vista and Windows 7, Linux kernel versions 2.6.35.X and Mac OS version OSX 10.6. There is though no particular feature of the developed software modules that would prevent its porting also to mobile platforms, notably Android.

Figure 2.9 depicts the deployment architecture for a peer containing CoMid and CoComp levels.

At the CoMid level we can identify Protocol and/or Technology Engines. At the CoComp level, CONVERGENCE supports CoSec (through the use of smart cards) and CoNet. Given that the first full-functional implementation of CoNet was available only for Linux, CoMid includes a CoNet TE to provide an abstraction layer for CoNet. Moreover, it is possible to have different implementations of the CoComp, (notably an implementation that still employs the address-based IP protocol stack currently in use on the Internet). Accordingly, the CoMid level is able to operate without the CoComp modules, operating directly on top of existing TCP/IP protocol stack. Likewise, the CoComp might be able to offer network services to other middleware frameworks.

The hardware and connectivity requirements of a full CONVERGENCE peer depend on the number of services it has to provide and the frequency of the incoming requests. As such, it cannot be deterministically calculated. Based on the requirements of the VM, and of the application servers, we estimate that the minimum requirements for a normal configuration are:

- CPU: >=2 cores, >=2.1 GHz/core;
- RAM: >=4 GB;

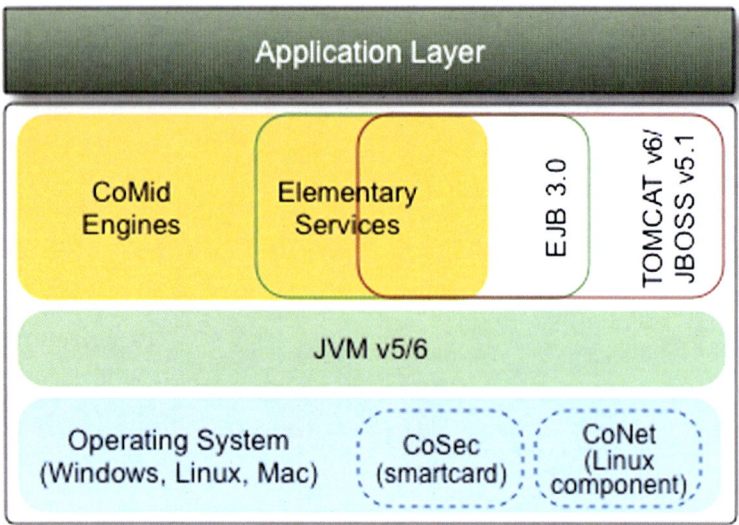

Fig. 2.9 Peer deployment architecture

• Storage: >=10 GB (minimum OS configuration).

However, configurations that support only a small number of Technology Engines may have lower requirements. If a peer contains only the Security TE and, to perform other operations, relies on remotely installed elementary services which it invokes whenever necessary, it may well run on a low cost laptop or a mobile device. All that is required is that it contains the software to run the local engines (JVM, libraries, etc.).

In addition to implementing the CoMid level, and optionally the CoApp level, it is mandatory that peers implement the CoComp level. As the other type of CONVERGENCE devices, the nodes, implement only the CoComp level, deployment details concerning this level are provided in the next section, dedicated to the description of the functionality and deployment architecture of CONVERGENCE nodes.

2.2.4.2 Node Deployment Architecture

Nodes are CONVERGENCE devices that implement CoComp level functionality. As already indicated, nodes may include the two CoComp components, CoNet and CoSec. Alternatively, they may contain only one of such components.

Figures 2.10, 2.11, 2.12, 2.13 and 2.14 present the functionality offered by each type of CoNet node.

Fig. 2.10 Functionality of border nodes

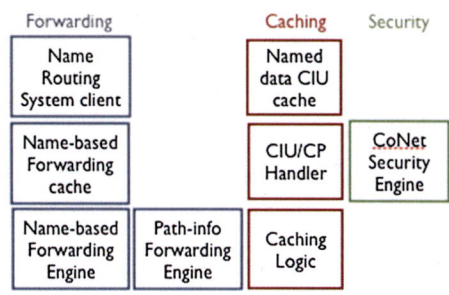

Fig. 2.11 Functionality of internal nodes

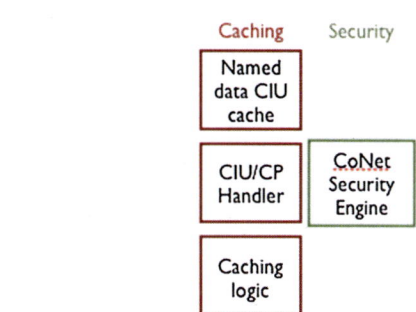

Fig. 2.12 Functionality of serving nodes

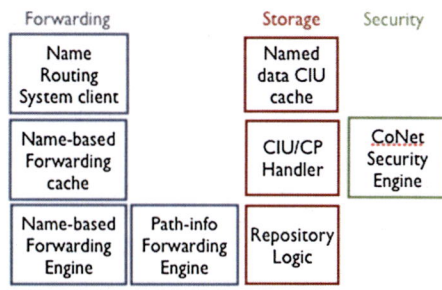

Fig. 2.13 Functionality of end nodes

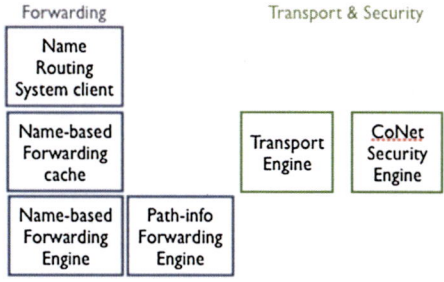

Fig. 2.14 Functionality of
NRS nodes

Name-based Routing Protocol	Routing Database	Name-based Routing System Server

2.3 Conclusions

The ultimate goal of CONVERGENCE is to facilitate and enhance the access to and transaction of resources in loosely coupled networked environments, such as the Internet. Resources can be media contents, services or the digital representation of real-world objects and people. CONVERGENCE has defined a common container, the Versatile Digital Item, based on an open standardized format, the MPEG-21 DI. This data unit embeds all information required to attain the above stated objectives. CONVERGENCE proposes the employment of this extended concept, together with a new layered architecture, to achieve a new operational paradigm for the Internet, whereby resources are routed, located and accessed based on their names and contents. The CONVERGENCE architecture implements the necessary functionality to efficiently use VDIs.

The defined architecture, structured in layers of functionality, is especially suited to enable multiple progressively evolving scenarios to be deployed on the Internet, from the current host-based Internet model towards the future Information-centric Internet. The definition of clear and open interfaces between such layers, guarantees up to some extent, independence between them, whilst offering developers the flexibility to use different and restricted sets of functionality as needed to satisfy distinct specific requirements. Each CONVERGENCE architectural layer comprises: sets of APIs, through which its services may be reached, the corresponding WSDL definitions and JAVA implementations; services to create, process and manipulate VDIs in networked and shared environments. These services can be used to provide specific functionality to upper layer applications, involving the exchange and transaction of digital resources or of digital representations of physical resources on the Internet. Third parties can develop varying applications supporting different business models by using the lower levels functionality as needed. Through those applications, users may benefit from increased quality of experience, gaining access to useful resources in a seamless way.

References

A. Detti, N. Blefari Melazzi, S. Salsano, M. Pomposini, "CONET: A Content Centric Inter-Networking Architecture", ACM SIGCOMM Workshop on Information-Centric Networking (ICN 2011), August 19, 2011, Toronto, Canada.

E. Aitenbichler, A. Behring, D. Bradley, T. Strufe, "Shaping the Future Internet", Ubiquitous Computing and Communication Journal, vol. 5, issue 2, 2010.

MPEG-21: ISO/IEC FDIS 21000-2:2005(E) MPEG-21 – Part 2: Digital Item Declaration, ISO, 2005.

MPEG-M: ISO/IEC 23006-1, Information technology—Multimedia service platform technologies—Part 1: Architecture, 2013a.

MPEG-M: ISO/IEC 23006-2, Information technology—Multimedia service platform technologies—Part 2: MPEG Extensible Middleware (MXM) API, 2013b.

MPEG-M: ISO/IEC 23006-3, Information technology—Multimedia service platform technologies—Part 4: Elementary Services, 2013c.

P.T. Eugster, P.A. Felber, R. Guerraoui, A. Kermarrec, "The Many Faces of Publish-Subscribe". ACM Computing Surveys (CSUR), vol. 35, issue 2, 2003.

T. Zahariadis, P. Daras, J. Bouwen, N. Niebert, D. Griffin, Aitenbichler F. Alvarez, G. Camarillo, "Towards a Content-Centric Internet", In: G. Tselentis et al. (Eds.) Towards the Future Internet - A European Research Perspective, IOS Press, 2010.

Chapter 3
The Network Level (CONET)

Andrea Detti, Stefano Salsano and Nicola Blefari Melazzi

Abstract CONET (Convergence Network) is the network-layer of the CONVERGENCE project. It is an Information Centric Network, which extends the CCNx one in several aspects, including routing scalability, transport mechanisms, security handling, integration with IP, etc. This section describes services and functionalities of CONET and reports some performance evaluations, carried out through laboratory and PlanetLab test-beds.

3.1 Introduction

Network level functionalities of the CONVERGENCE system are provided by an Information-Centric Network (ICN), named CONET (Convergence Network).

The ICN paradigm envisages a network-layer thoroughly meant for information dissemination, rather than for point-to-point transfers of raw bits (Cheriton and Gritter 2000; Koponen et al. 2007; Jacobson et al. 2009a; Smetters and Jacobson 2009; Trossen et al. 2010; Detti et al. 2011a). In ICN, the network layer provides users with named content, instead of communication channels between hosts. The basic functions of an ICN infrastructure are to: (i) address content by adopting an addressing framework based on names, without a reference to the current content location (i.e., location-independent names); (ii) route a user request, based only on the content-name, towards the "closest" location containing the required content; potential locations include not only the origin server of that content but also network caches or even devices of other users that downloaded the same content beforehand; (iii) deliver the content back to the requesting host.

A. Detti (✉) · S. Salsano · N. Blefari Melazzi
Electronic Engineering Department, University of Rome Tor Vergata,
Via del Politecnico 1, 00133 Rome, Italy
e-mail: andrea.detti@uniroma2.it
URL: http://netgroup.uniroma2.it/people/faculties/andrea-detti/

F. Almeida et al. (eds.), *Enhancing the Internet with the CONVERGENCE System*, 49
Signals and Communication Technology, DOI: 10.1007/978-1-4471-5373-3_3,
© Springer-Verlag London 2014

The main ICN benefit is a simplification of design, deployment and management of content distribution services. In this sense, an ICN does not extend the set of end-user services that an IP stack could provide; rather simplifies the business of content and service providers.

Currently, major content and service providers "patch" the inefficiency of IP data dissemination, by using dozens of custom extra-IP functionalities, e.g. HTTP proxies, Content Delivery Networks, multi-homing, multicast delivery, etc. The drawback of such heterogeneous deployment is the burden of achieving an efficient interplay among several functionalities, in case offered by different company. This tune up is critical and complex, not only from a technical point of view but also from a management one.

An ICN is directly meant for information dissemination. Consequently, an ICN relieves providers from arranging extra-ICN functionalities. For instance data replication, caching, multi-homing and multicast delivery are inner ICN functionalities, directly handled by the network layer, so simplifying network design, deployment and management.

Through an ICN Application Programing Interface (API), a user may request by-name an information item and the ICN network-layer provides the user with the item, fetching it from a serving device that the network-layer has autonomously selected. Serving device could be the original server publishing the information, a replica of the original server or a cache of a network device or even a device of another user that downloaded the same information item beforehand.

Through an ICN API a content-provider may publish by-name an information item and the ICN network-layer properly configures its routing plane so that requests of the published item will be served.

An ICN is aware of distributed information, since the network-layer identifies each information item with a unique name-based identifier, and uses data units that include such name-based identifiers. ICN data-units may also convey request of information items. In both cases, a node is aware of "what" a data-unit refers to. This awareness is the enabler of content-based functionalities such as: (i) routing-by-name, (ii) caching; (iii) support for mobile, multicast and peer-to-peer communications; (iv) support for time/space-decoupled model of communications; (v) content-oriented security model; (vi) content-oriented quality of service, access control and traffic engineering.

From theory to practice, the design of an ICN architecture sets up several technical challenges, as for instance:

Primitives & interfaces—define the relationship of the ICN protocols with the overall architecture, including their positions and connections with IP and current protocols; for instance, will ICN be the new narrow waist of the Internet (i.e. the lowest-level global network primitive, the only way to establish global communication) or will it sits over IP? Other important issues of this component include the basic primitives and interfaces, e.g., does the ICN layer offer a publish/subscribe primitive or is this functionality delegated to upper layers? How can we change/extend the socket API, which is one of the main responsible of the current "ossification" of the Internet?

Naming scheme—the naming-scheme specifies the identifiers for the information addressed by the ICN. The choice of a naming-scheme impacts different aspects an ICN, including the handling of name uniqueness and trademarking, routing scalability, flexibility of supporting different applications, usability, security. For instance, human-readable flat-names improve usability; hierarchical names foster aggregation of names in name-based forwarding tables; names containing the public key of the owner of the content simplify security.

Name resolution—a node routes by-name a request of an information item towards a selected serving device. Hence, the ICN node should resolve the name of requested items in the "physical" address of next ICN node towards the selected serving device. Several name-resolution approaches are possible, ranging from an off-path resolution, e.g. based on DHT, to an en-route hop-by-hop resolution exploiting name-based routing tables.

Routing scalability—with respect to IP, the routing plane of an ICN has to handle a number of information items and corresponding names that is much bigger than the number of IP network prefixes. This has implications on the size of ICN routing tables, on the complexity of lookup functions and on the distribution of ICN routing information and is one of the main concerns of ICN.

Information delivery—an ICN addresses information items rather than hosts, so it needs a technique to route back the requested information item from the serving device to the requesting device. Some architectures propose to use plain IP means, other ones propose to face the issue through ICN's own means.

Segmentation mechanisms—these mechanisms are needed to split an information item or a content in different chunks (each chunk is an autonomous data unit with embedded security and addressable by the routing plane). Content to be transported over an ICN can be very variable in size, from few bytes to hundreds of Gigabytes. Therefore it needs to be segmented in smaller size data units, typically called chunks, in order to be handled by ICN nodes. A chunk is the basic data unit to which caching and security is applied.

Transport mechanisms—as regards the transport protocol, we favour a receiver-driven approach: in ICN the transport of an information item does not exploit an end-to-end session and while the requesting device remains the same, the serving device may change also on a chunk-by-chunk basis. This requires a complete rethinking of actual Internet transport mechanisms towards a receiver-driven approach, where the whole transport logic is on the receiver side. Serving applications split information items in chunks and assign unique names to chunks. On the other side, application clients fetch sequentially these chunks, according to a receiver-driven transport algorithm. As in IP, transport mechanisms should be tailored to application characteristic, e.g. file download, video streaming, voice over ICN, etc.

In-network caching—ICN nodes may cache chunks of information. Differently from traditional HTTP caching, an ICN is a *cache network*; this implies the need of properly devising a replication strategy that optimizes the caching space, e.g. by avoiding excessive duplication of content in the network caches.

Security and privacy challenges—security and privacy issues in ICN tackle several aspects: (i) integrity: received content has not been modified, i.e. it is the originally published one; (ii) provenance: source of the content is authentic, i.e. the data is provided by the original creator; (iii) relevance: received content is really the content requested by the user. The verification of these criteria should be done not only at the receiver side, but also in network nodes, as it is important that the network be protected from pollution of content-caches with fake information items. In addition, the network should protect information consumers from profiling or censorship of their requests. These issues, extensively addressed in traditional networks, require a significant rethinking when challenged against the unique distinguishing characteristics of ICNs. Traditional network security protocols such as IPsec or TLS focus on protecting the communication between an information consumer and a content server, and do this by deploying trustworthy infrastructures devised to enforce authentication and access control primitives on dedicated servers. In ICN, the requested content is not anymore associated to a trusted server or an endpoint location, but it can be retrieved from, say, a network cache managed by an hardly trusted administrative domain. This calls for data-centric security and privacy solutions, being hardly viable a secure infrastructure which involves storage servers and network caches in heterogeneous non-collaborative domains. The data-centric security model increases the communications overhead with respect to traditional IP-based solutions, where security related information are exchanged only one time per information transfer, i.e. at the start of the end-to-end session. Furthermore, the architectural binding of security and network layer functionality has to be carefully designed. In fact, it may impose severe deployment limitations. For instance, an ICN architecture based on digital certificate may not properly operate without a PKI, thus making difficult to realize self-forming ICN networks. Another example regards the denial of service due to the presence of fake content in the cache, caching should be secure, i.e. node must verify the validity of cached contents. This operation should be carefully devised to operate as much as possible with a time scale close to the line rate of nodes. Otherwise only a limited set of forwarded information items can be cached by a node.

Push services—an ICN is primarily meant to enable clients to "pull" information items. However, today, several Internet services are customized to the user and these services require that client "pushes" information to server. For instance home banking, trading on-line, dynamic web servers, belong to this class of services. Therefore, an ICN architecture aiming at being the narrow waist of the Future Internet has to support also a push service model.

Smooth migration path—an ICN architecture should be usable and deployable in a scalable way, i.e. should support a smooth migration path from current applications and networks technology based on TCP/IP, to ICN applications and networks.

In this section, we describe the Information Centric Network of the Convergence project, namely CONET (Detti et al. 2011a, 2012b). In short, we can say that CONET:

1. Uses a hierarchical naming scheme;
2. Performs name-resolution on-path, by using name-based routing table;
3. Copes with the routing scalability issue by using our proposed Lookup and Cache routing architecture;
4. Delivers information to requesting user either by means of control data temporary left in network nodes, or inserted in network data units;
5. Implements an efficient receiver driven TCP-like transport algorithm for information delivery services;
6. Uses Identity Base Signature and/or self-certifying names to carry out security related operations;
7. Provides in-network secure caching;
8. Supports push services by our proposed named-sap concept;
9. Can be deployed with an IP overlay or a clean slate approach or with our proposed integration approach (see Detti et al. 2011a, b).

Some of these features (1, 2, and 4) are derived from the CCNx architecture (Jacobson et al. 2009b); the remaining ones are CONET's own. In what follows, we describe CONET in details.

3.2 Integration of the CONET Network Layer in the CONVERGENCE system

Figure 3.1 describes the architecture of the CONVERGENCE system. We have publishers that wish to provide customers with their information or services (e.g. text, picture, movie, home bank access, etc.). Information or services are described by a set of metadata that are embedded in the *middleware* data-unit, namely the VDI. The VDI can either contain also the actual information item or contain only a reference to it, where the reference is a network-layer identifier.

VDIs are exploited by middleware *engines* for different aims; engines are middleware technologies/protocols carrying out specific tasks. For instance, the middleware can exploit VDIs to offer content-based publish-subscribe services (Chiariglione et al. 2012; Eugster et al. 2003) as follows: applications express their interests; the interest is stored by the middleware and when a corresponding information item or service is published, the middleware returns the matching VDIs; at this point, the application can exploit the CONET to either fetch the desired information item or interact with a server providing the desired service; this is possible since the VDI contains the network-layer reference to the information item or to the service.

However, publish-subscribe is not the only possible interaction model. For instance, an application could also use a request-response model in which a search engine provides the user with the VDI matching her request.

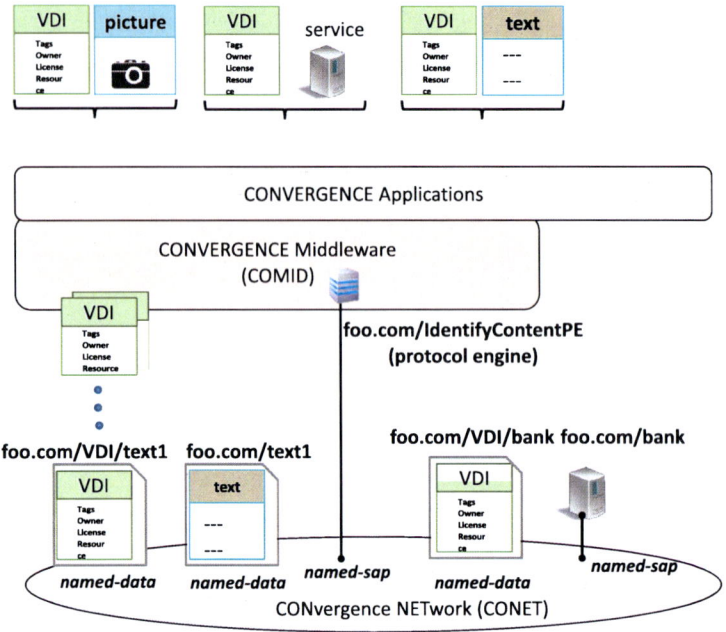

Fig. 3.1 Integration of CONET within the CONVERGENCE system

The VDI and the related information item or service are accessed through the CONET as *named-resources*. A named-resource can either be an information item or a service, addressed by the CONET by means of a NEtwork Identidier (NID), which is a name.

Named-resources that refer to information items are called *named-data*. Figure 3.1 reports a case where a text file and its associated VDI are disseminated in the CONET as two named-data items. The named-data item related to the text file has the name (i.e., a NID) "foo.com/text1", and the named-data item related to the VDI has the name "foo.com/VDI/text1".

Named-resources that refer to services are called named-service-access-point, briefly *named-sap*. A named-sap is merely a "port" towards an upper layer entity addressed by the CONET through a name. The use of named-saps enables CONET to support *interactive* services, i.e. services that require a custom client/server interaction to provide personalized contents or actions. Examples of this interactive service class include: services offered by Protocol Engines (PEs) of the CONVERGENCE middleware (COMID), HTTP servers of dynamic Web pages, home banking or trading on-line HTTP servers and SMTP servers.

Figure 3.1 reports also a case where a home banking service and its associated VDI are disseminated in the CONET as a named-sap and a named-data item, respectively. The named-sap of the home banking service has the name "foo.com/bank", and the named-data of the VDI has the name "foo.com/VDI/bank". The figure reports a named-sap associated with a COMID IdentifyContent PE, which is addressed by the CONET through the "foo.com/IdentifyContentPE" named-sap.

3.3 CONET Services

3.3.1 Publication of Named-Resources

The CONET enables users to publish and to revoke named-resources. A resource can be replicated in different geographical locations by using the same name.

Figure 3.2 reports the case of a user that publishes both a named-data item and a named-sap. The named-data of Fig. 3.2 is a text file identified by the CONET with the name "foo.com/text1". To publish the named-data, the user exploits the API provided by a CONET Serving Node (SN) to *store* the named-data item in a local Repository, and to *advertise* the presence of "foo.com/text1" on the CONET routing plane.

The named-sap of Fig. 3.2 is a home banking service provided by a Server and identified by the CONET with the name "foo.com/bank". To publish the named-sap, the user exploits the API of the Serving Node to *advertise* the reachability of the service "foo.com/bank" on the CONET routing plane.

Albeit not reported in Fig. 3.2, the user may exploit the CONET API to remove the text file from the repository and to withdraw the identifier "foo.com/text1" from the CONET routing plane. A similar action may occur in case of revocation of the home banking named-sap.

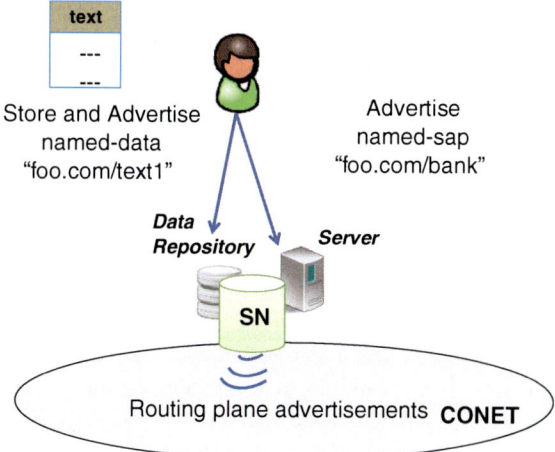

Fig. 3.2 Publication of named-resources

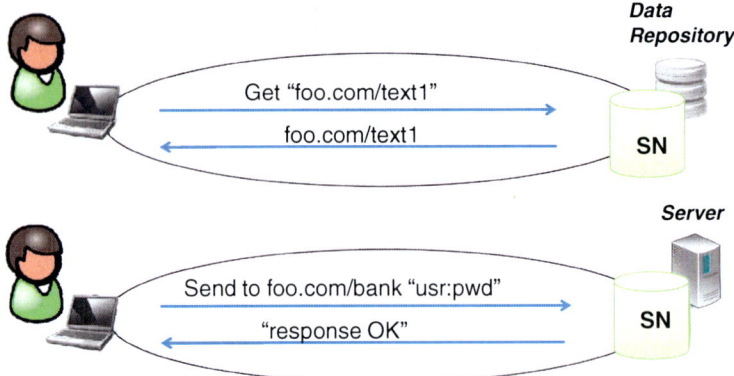

Fig. 3.3 Access to named-data and named-sap

3.3.2 Access to Named-Resources

CONET provides users with the possibility of accessing named-resources by using their network identifiers (NIDs), i.e. their names. When a named-resource is a named-data item, the CONET delivers it to intended recipients. When a named-resource is a named-sap, the CONET provides the means to exchange information between a requesting upper layer entity and the upper layer entity addressed by the named-sap.

Figure 3.3 depicts the access to named-resources provided by the CONET. In case of the named-data item "foo.com/text1", the network *routes-by-name* the request of "foo.com/text1" towards the best Serving Node that has the item. The Serving Node provides the named-data item, and the CONET sends it to the recipient. As we will see in the next section, a CONET node may cache named-data items, therefore the named-data "foo.com/text1" item could also be sent to the recipient by an en-route node, rather than from the original Server.

Figure 3.3 also reports the access to a named-sap. In this example, the user sends her credential ("usr:pwd") towards a remote named-sap, whose name is "foo.com/bank" and gets back a response from the Server.

3.4 Naming Model

To identify named-resources, the CONET uses a hierarchical naming scheme, formed by a *PrincipalId* and a *Label*, i.e. *PrincipalId/Label*. PrincipalId is a string, e.g. "foo.com", that uniquely identifies the principal of the resource. Label, e.g. "text1", is an identifier that uniquely identifies a resource among those published by a principal. To make a comparison with Internet URL, the PrincipalId is the domain-name and the Label is the Path.

In general, both the PrincipalId and the Label are formed by name-components that are strings separated by the "/" character. For instance, the identifier "foo.com/content/doc/text1" is formed by foo.com as PrincipalId and "content/text/text1.txt" as Label, and this Label has three components: "content", "doc", and "text1".

A PrincipalId could be a human-readable name or the public key of the principal. In the latter case, we have a so called *self-certifying name* (Koponen et al. 2007). Therefore CONET supports both human-readable and self-certifying names.

3.5 Data Model

3.5.1 Data Model for Named-Data Related Services

Figure 3.4 reports the data model used by the CONET for named-data. An information item is linked with a unique network identifier, e.g. "foo.com/text1", so as to form a named-data.

A named-data is segmented in different chunks and each chunk is packaged in a data unit called *named-data* Content Information Unit (CIU). A named-data CIU is uniquely identified by the network with the name of the whole named-data item and an identifier of the chunk number. The identifier of the chunk number can either be included in the name, e.g. "foo.com/text1.txt/chunk1", or included in a suitable field of the CIU header (Detti et al. 2011a). A named-data CIU contains security information (Smetters and Jacobson 2009), so that traversed nodes may securely cache it, by avoiding denial of services attacks due to caching of fake contents. Thus, the named-data CIU is the caching data-unit of the CONET.

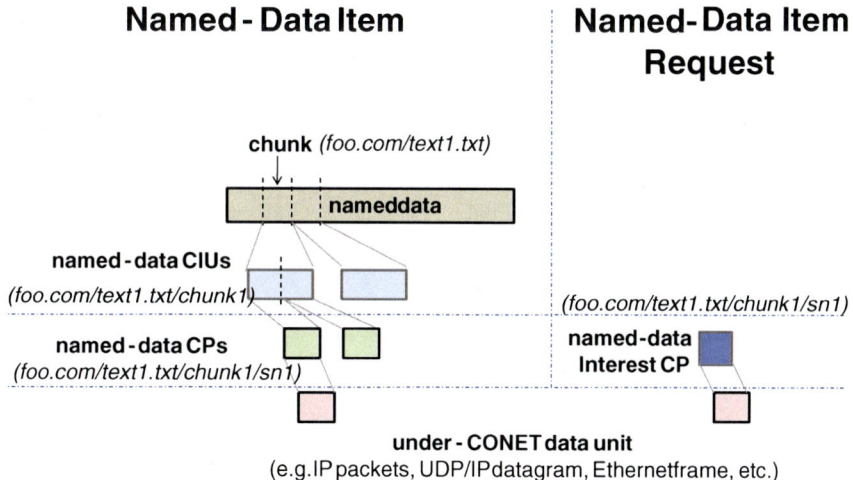

Fig. 3.4 Data model for named-data services

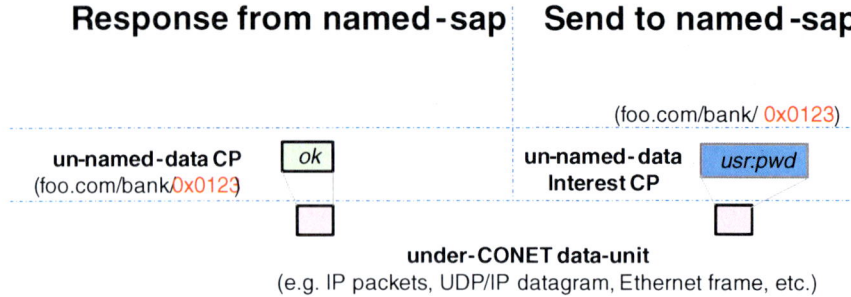

Fig. 3.5 Data model for named-sap services

To reduce the security overhead and the rate of security checks performed by caching nodes, the size of the named-data CIU should be greater than the usual payload transported by IP packets. For instance, reasonable size of the named-data CIU could be in the order of tens or hundreds of IP packets/Ethernet frames. For this reason, a named-data CIU is further segmented into so-called named-data *carrier packets* (CP), whose size fits the end-to-end maximum transfer unit. A named-data CP is uniquely identified by the network with the name of the named-data CIU plus a segment number; also in this case, segment number can either be included in the name, e.g. "foo.com/text1.txt/chunk1/sn1", or included in a suitable field of the CP header (Detti et al. 2011a).

As we will see in the next section, applications download a named-data by sequentially downloading all CIUs and, hence, all CPs. To request a named-data CP, a user issues a named-data Interest CP,[1] which includes the identifier of the desired carrier packet (see Fig. 3.4).

3.5.2 Data Model for Named-Sap Related Services

Figure 3.5 reports the data model used by the CONET for named-sap related services.

A named-sap is the coupling between a server upper layer entity and a network identifier, i.e. a name. For instance Foo's server for home banking could be addressed by the CONET through the named-sap "foo.com/bank".

A client upper layer entity interacts with a server upper-layer entity by exchanging data that are of exclusive interest of such client–server interaction and, hence, are *un*-cacheable for further re-use. For this reason, we refer to this type of data as *un*-named-data.

---

[1] We note that in (Detti et al. 2011a), we denoted this named-data Interest CPs with another name: Interest CIU. We changed the name because an Interest CIU is directly mapped in an underlying carrier-packet.

Client data (e.g., "usr:pwd") are sent to server through data units called un-named-data Interest CPs. These messages include the network identifier of the named-sap and a nonce, used to identify the specific *request-response* interaction; e.g. "foo.com/bank/0x0123", where "foo.com/bank" is the identifier of the named-sap and "0x0123" is the nonce.

Server responses are sent back within un-named-data CP, which includes the network identifier of the named-sap and the nonce of the Interest CP.

3.6 Service Interaction Model

This section describes the CONET service interaction model for named-data and named-sap related services, respectively.

3.6.1 Interaction Model for Named-Data Pull Services

Figure 3.6 (left) depicts the end-to-end interaction for the fetching of a named-data. An end-node downloads a named data by retrieving all its named-data CIUs and, hence, CPs. To download a named-data CPs, a client sends out a named-data Interest CP (briefly, 'data CP' in the figure), which contains the identifier of the

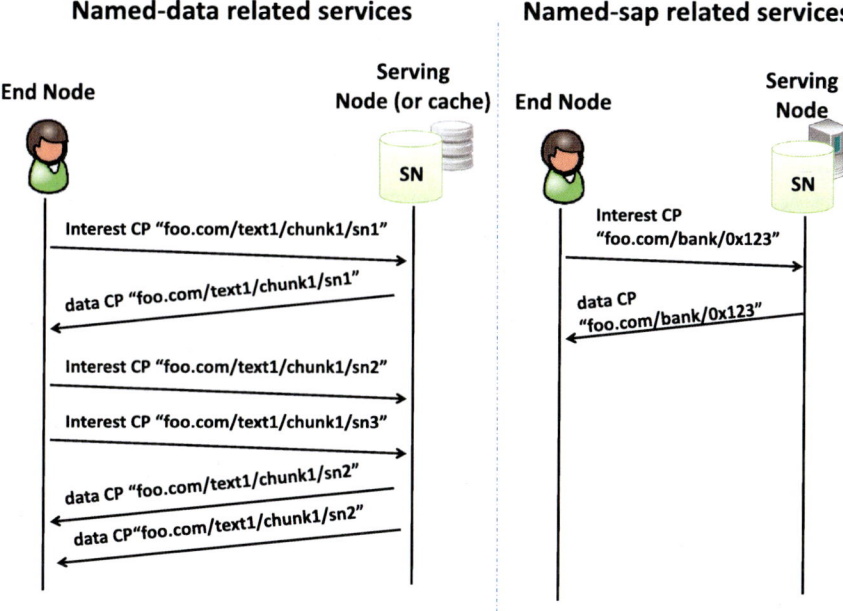

Fig. 3.6 End-to-end interaction models for named-data (*left*) and named-sap (*right*) services

desired named-data CP, e.g. "foo.com/text1/chunk1/sn1". The CONET routes-by-name the Interest message toward the Serving Node, which sends back the requested data within a named-data CP (briefly, 'data CP' in the figure). To download the whole set of named-data CPs, the client may adopt a TCP-like receiver-driven approach (Salsano et al. 2012; Kuzmanovic and Knightly 2007), working as follows: (i) TCP ACKs are replaced by Interest CPs; (ii) TCP segments are replaced by named-data CPs; (iii) traffic control operation is carried out at the receiver side, by using a TCP-like congestion window (cwnd) control, applied on the number of in-flight Interest CP messages.

We observe that the use of a receiver-driven TCP-like transport algorithm is suitable for download services, while in the case of e.g. live streaming (Detti et al. 2012c) or voice (Jacobson et al. 2009b), different receiver-driven flow control mechanisms should be designed, according to the specific application requirements.

3.6.2 Interaction Model for Named-Sap Push Services

Figure 3.6 (right) depicts the end-to-end interaction between an end-node and a remote server behind a named-sap. In this case the interaction model is a request-response, as previously described in Sect. 3.5.2.

3.7 Architecture and Functions

Figure 3.7 reports the CONET network architecture. CONET nodes are inter-connected by "sub-systems" that can be implemented in several different ways. For instance, a sub-system could be a public or private IP network, an overlay UDP/IP link, a layer-2 network, a PPP link, etc. This is the same concept used in current IP networks, in which IP hosts and routers can be connected via different layer 2 technologies. A CONET sub-system may include: (i) CONET end-nodes that access named-resources; (ii) CONET Serving-nodes that provide named-resources and (iii) CONET nodes that relay carrier-packets between sub-systems and optionally cache named-data CIUs.

Routing (by-name) of Interest—To route an Interest CP, a generic CONET node uses a name-based Forwarding Information Base (FIB), whose entries are in the form < name-prefix, next hops > . A longest matching algorithm, based on characters, selects the best entry of the FIB. For instance, in case of Fig. 3.7, when node N1 receives an Interest CP for "foo.com/text1/chunk1/sn1", then the longest match selects the FIB entry "foo.com" (see the FIB reported in the top of the figure). Then node N1 forwards the Interest CP towards the next CONET interface of the path, i.e. interface #C of N2.

The whole set of possible name-prefixes could be too large to be stored in the "fast" FIB memory, e.g. based on SRAM or TCAM technology. To overcome this

Forwarding Information Base (FIB) of N1

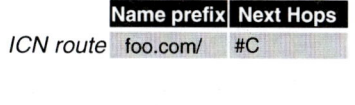

Name prefix	Next Hops
ICN route foo.com/	#C

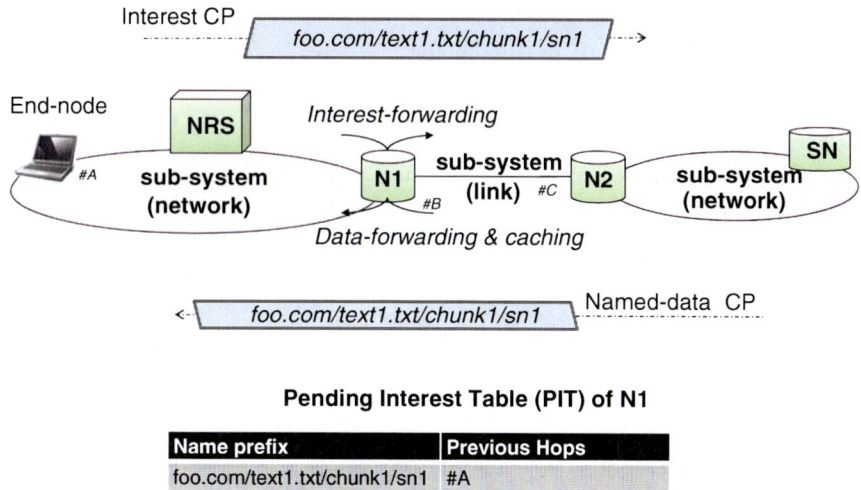

Pending Interest Table (PIT) of N1

Name prefix	Previous Hops
foo.com/text1.txt/chunk1/sn1	#A

Fig. 3.7 Network architecture

aspect of routing scalability, we propose that (Detti et al. 2012a; Blefari Melazzi et al. 2012b):

- The PrincipalId is the *longest* routing prefix; i.e. name-prefixes contained in the routing table distributed at the inter-domain level either are PrincipalIds or are shorten than PrincipalId, in order to aggregate more PrincipalIds in a same routing entry. This rule has to be applied at least at the inter-subsystems (or inter-domain) level;
- The FIB is used as a *route cache* (Changhoon et al. 2009), to temporarily store the limited set of routes necessary to support the on-going end-to-end interactions. We call these routes *active routes*.

On the base of real Internet traces and in a case where PrincipalIds are the actual 2×10^8 Web domain-names, in (Detti et al. 2012b) we show that the set of active routes is relatively small (10^4), both with respect to the whole set of possible routes (2×10^8), and also with respect to the current storage capacity of SRAM memory technology (10^6 name prefixes) (Zhao et al. 2010; Perino and Varvello 2011).

In case the FIB does not contain an entry to route-by-name an incoming Interest CP, the node lookups the routing entry in a Routing Information Base (RIB), deployed in a centralized Name Routing System (NRS) node, which hosts the

CONET Routing Engine. A RIB may be implemented through a bank of "slow" DRAM memories, whose overall memory space is able to store the whole set of possible name-prefixes. We call such routing-by-name architecture *Lookup-and-Cache*.

Figure 3.8 depicts an example of Lookup and Cache operations during the forwarding of an Interest CP. When node N1 receives the Interest CP for "foo.-com/text1/chunk1/sn1", the node does not have a valid routing entry in the FIB, hence lookups the entry in the remote RIB. Then it inserts the routing entry ("foo.com") in the FIB and forwards the Interest CP.

We observe that the Lookup and Cache architecture is perfectly in line with the Software Defined Network paradigm: the NRS node functionality could be implemented in the central network controller (McKeown et al. 2008; Blefari-Melazzi et al. 2012a).

Routing (by-name) of Data—To support *information delivery,* i.e. to route data from serving node to the requesting device, a possible, *state-full*, approach is that a node temporarily stores the couple < CP identifier, previous-hop interface list > in a Pending Information Table (PIT), during the forwarding of an Interest CP. The PIT contains information about the set of Interest CPs received by a node and not yet served, i.e. messages for which the node has not yet sent back the related named or un-named CPs. PIT entries are grouped by name and an interface list field contains the addresses of the previous-hop interfaces that forwarded the related Interest messages. For instance, in the case of Fig. 3.7, when node N1 receives the Interest "foo.com/text1/chunk1/sn1" forwarded by the previous end-node, then node N1 inserts in the PIT the entry < "foo.com/text1/chunk1/sn1", #A>, where #A is the physical address (e.g. IP or Ethernet address) of the end-node interface.

When a node receives a named/un-named data CP, it lookups the enclosed name in the PIT, forwards the CP towards all the interfaces contained in the interface list and deletes the PIT entry. In this example, when node 1 receives a named-data CP for "foo.com/text1/chunk1/sn1", then it lookups the entry in the PIT, forwards the message towards the #A interface and deletes the PIT entry.

In-network, en-route, caching—A CONET node may cache received named-data CIUs in a local memory. Since a named-data CIU is an aggregation of named-data CPs, then caching involves a reassembly operation. Moreover, a CIU is inserted in the cache only if the enclosed security information confirms its proper validity. When an en-route node receives a named-data Interest CP, the node first checks the presence of the related named-data CP in its cache. If a cache hit occurs, the node directly sends back the named-data CP. In case of cache miss, the node executes the forwarding and PIT operations previously described.

In addition to en-route caching, other caching approach could be used, depending on the specific environment. For instance, in (Gallucio et al. 2012; Detti et al. 2012b) authors propose a caching approach based on overhearing, for satellite networks.

Routing protocol—The lookup and cache routing architecture requires a routing protocol to distribute name-prefixes and setup the RIBs of NRS nodes. The

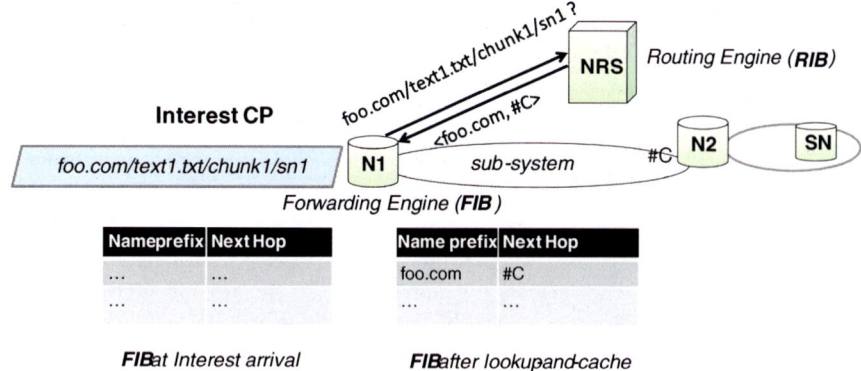

Fig. 3.8 Lookup and Cache routing-by-name architecture, Interest forwarding

Lookup-and-Cache architecture is independent from the specific routing protocol implementation. To show an example, we implemented a simple routing protocol based on the REGISTER and UNREGISTER functions proposed by the DONA architecture (Koponen et al. 2007), adapted to our specific network model.

As show in Fig. 3.9, a Serving Node (SN) REGISTERs and UNREGISTERs the name-prefixes (e.g., "foo.com") of "its" named-data items in the local NRS node. The local NRS node and the next ones forward the REGISTER/UNREG-ISTER messages toward their parents and peers neighbours up to a NRS of the tier-1 level. As in the case of an Interest message, the routing of REGISTER/UNREGISTER messages is by-name. Indeed, each NRS node has a named-sap used to receive routing messages, e.g. "ss1.org/routing.sap". Reception of REG-ISTER and UNREGISTER messages enables an NRS node to properly setup its RIB. We observe that the same NRS may serve more than one sub-systems or nodes and even a whole autonomous system (i.e. a collection of subsystems administered by the same entity, as in the Internet). In this case, the NRS would have a RIB for each served node.

3.8 Security Support

CONET security refers to the ability of a node to verify the validity (integrity, provenance and relevance, see Sect. 3.1) of named-data CIUs. The way security functionality is designed and implemented depends on the adopted naming scheme: human-readable or self-certifying names (see Sect. 3.4). Moreover, in order to support in-network secure caching (Ghodsi et al. 2011), named-data CIU should contain all the security information needed to verify its validity, i.e. the signature and the means to verify the signature, such as the public key of the signing principal (Detti et al. 2013).

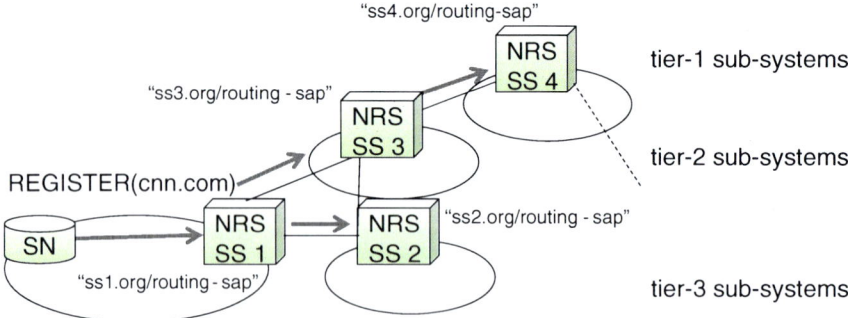

Fig. 3.9 Lookup and Cache architecture: inter-domain distribution of name-prefixes

In case of human-readable names, the CONET needs a centralized authority to control PrincipalId uniqueness. Moreover, the CONET needs a PKI infrastructure to control validity of principal's public key. To reduce the bandwidth overhead of security, we propose to use Identity Based Signature approach, where the public key is just the PrincipalId, already contained in the name of the named-data CIU.

In case of self-certifying names, PrincipalId is the principal public key that could be randomly generated by the principal. Moreover, in this case, security related verifications do not require a PKI, as discussed in (Ghodsi et al. 2011; Detti et al. 2013). Therefore, we argue that self-certifying names are a good choice for self-forming ICNs, where the presence of a PKI may be impractical.

3.9 Migration Path

The migration from services based on the TCP/IP API to services based on the CONET API requires, on the one hand, to have applications able to operate on the new API and, on the other hand, to deploy the CONET.

Regarding applications, we observe that the CONET API is quite similar to the HTTP API, which is the most used API by Web applications (Popa et al. 2010). This similarity eases the development of HTTP/CONET (or generally ICN) transparent proxies, which can be used during the migration from HTTP based application to CONET based applications. For instance, we followed this approach to design a video streaming application (Detti et al. 2012c), in which the video client (VLC) is a plain HTTP application connected to a HTTP-to-CONET proxy.

Regarding the deployment of the network, we envisage a possible migration path which starts from a first *overlay* deployment (similar to CCNx (CCNx 2012)) where CONET uses IP as carrier and adopts dedicated hardware; to a second, more *integrated*, scenario where CONET and IP use the same data-units, and hence the same hardware.

Technically, in the first deployment scenario, CONET carrier-packets could be transported in the payload of UDP/IP packets, thus the connections among CONET nodes would be overlay links and CONET nodes would use specific hardware. In the second deployment scenario, control information of carrier-packets is integrated in the IP header, as an IPv4 options or IPv6 extension header (Detti et al. 2011a, b); in doing so, CONET nodes and IP routers could be integrated in a same hardware.

3.10 Performance Analysis

In this section, we report a brief survey of the performance analysis of some CONET aspects, namely routing and transport issues. A more detailed descriptions of measurements and performance figures are contained in referenced papers.

3.10.1 Routing

In (Detti et al. 2012b; Blefari-Melazzi et al. 2012b) we use real Internet traces to assess the feasibility of using the FIB as a cache of routes. We considered a deployment scenario where CONET is used to fetch current Web contents. In this case the name-prefixes handled by the CONET routing plane are the domain-names, which nowadays are in the order of 10^8. This amount is two orders of magnitude greater than the storage capacity provided by current FIBs; indeed, a SRAM memory may store a number of name-prefixes in the order of 10^6.

CONET copes with such storage limitation by using the FIB as a cache of *active* routes, i.e. of those name-based routing entries currently needed by a node to forward traffic. To prove that this approach is feasible we must show that the number of active routes is lower than the FIB storage capacity.

To verify this feasibility, we use a real Internet trace of a tier-1 Internet link (Equinix-sanjose-dirA) from which we derive a hypothetical CONET trace of a tier-1 node, as shown in (Blefari-Melazzi et al. 2012b). Then, we use the hypothetical trace to compute the number of CONET active-routes that a tier-1 node would have. Figure 3.10 reports the number of active-routes versus time and we observe that this number is in the order of 5×10^3, so much lower than the FIB capacity. Therefore we can state the using the FIB as a route-cache is feasible in the assumed scenario in which COINET is used to fetch current Web contents.

In (Detti et al. 2012b) we evaluate the possible performance degradation introduced by the lookup and cache routing operation, in the laboratory scenario depicted in Fig. 3.11. We have an ICN node between two subsystems. Within subsystem A, we have end-nodes (clients) that fetch named-data items from a serving node, located in subsystem B. The serving node publishes 10,000 named-data items and each named-data item has a different, random, name-prefix.

Fig. 3.10 Number of active-routes for the Equinix-sanjose-dirA trace

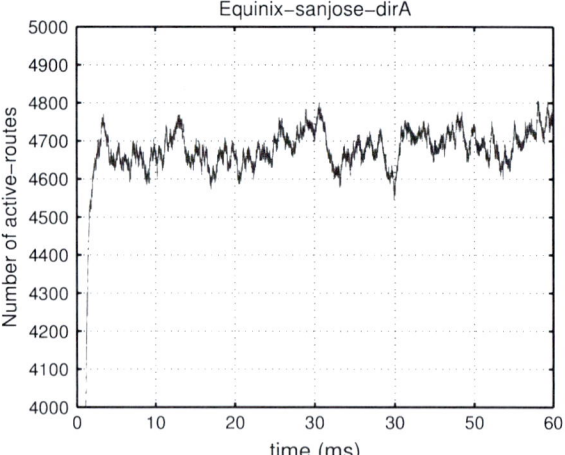

Fig. 3.11 Lookup and cache testbed setup

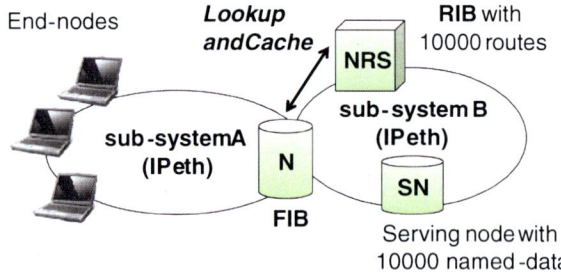

Consequently, in this scenario a full FIB would contain such 10,000 name-prefixes. On each node, we disabled in-network caching of named-data CIUs, to avoid the influence of data caching during the routing performance assessment.

Figure 3.12 reports the average download time of a named-data item versus the number of active routes. We consider three cases: in the first case we consider node N with a FIB of size 100 entries, which uses the LRU replacement policy to replace cached routes; in the second case we use another replacement policy based on a Inactivity Time Out (ITO) estimator (Detti et al. 2012b); in the third case we assume that node N has an unlimited FIB space, properly preloaded with the whole routing set (i.e. 10,000 routes). Results show that, as expected, when the number of active-routes is lower than the available FIB space, performance degradation introduced by a limited FIB is practically negligible, with respect to the case of unlimited FIB. Conversely, when the FIB is overloaded, i.e. the number of active-routes is greater than the FIB space, then performance degradation shows up and the ITO policy outperforms the simpler LRU.

In (Blefari-Melazzi et al. 2012b) and 2012c, we repeated the same download delay measurement on the PlaneLab *overlay* CONET deployment shown in Fig. 3.13, where links are UDP/IP sockets that transfer CONET carrier-packets.

Fig. 3.12 Average download time versus number of active-routes in laboratory test-bed

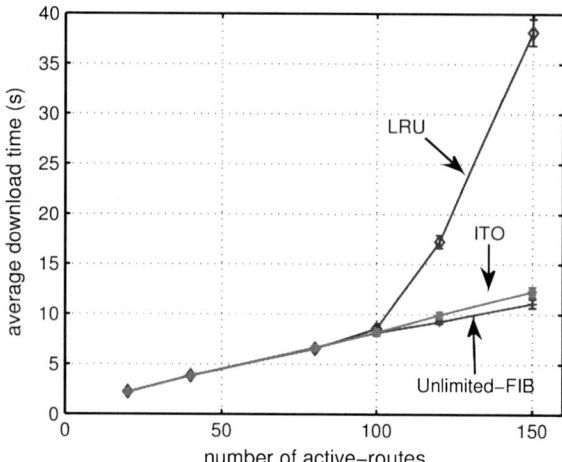

This scenario represents the case of a CONET *Autonomous System*, which is an aggregation of CONET subsystems under the control of the same network operator. The topology graph is generated so as to resemble the European GEANT research network. Each node is labelled with the country code name of its actual location, and serves a CONET subsystem, formed by a serving-node and an end-node (the latter is shown in the figure only in the case of the IE node). Serving-nodes publish a wide set of named-data items, whose popularity follows a Zipf distribution. End-nodes fetch named-data items, with exponential negative distributed inter-arrival times between two consecutive fetches. Lookup and Cache routing is supported by a single NRS node, which serves all nodes of the Autonomous System. Each node is connected to the NRS with a best-effort UDP/IP socket, used to perform routing lookup operations. This path is not drawn in Fig. 3.13.

Figure 3.14 shows the average download time versus the FIB size (number of entries), comparing the case of nodes without content (named-data) cache and the case of nodes with a content cache; the content cache size (number of entries) is equal to 10 % of the total number of published named-data items. The x-axis includes also an out-of-scale point, representative of a full preloaded FIB (labelled "Full-FIB") where, for each node, we use an *unlimited* FIB, pre-loaded with all name-based routes that the node could use. This measurement allows highlighting the worsening of performance deriving from the use of a limited FIB as a cache of routes and from the use of a centralized remote RIB, located in the NRS node.

As expected, as the FIB size increases, the performance tends to the full-FIB case, while caching contents leads to a decrease of the download time, since some named-data CIUs are delivered by the cache of nearby nodes, rather than from far away servers.

If we look at the curve representing the no-content cache case, the download time decreases of about 600 ms, when the FIB size increases from 50 to the full-FIB case. We argue that this *lookup delay penalty* is due to the connectivity/

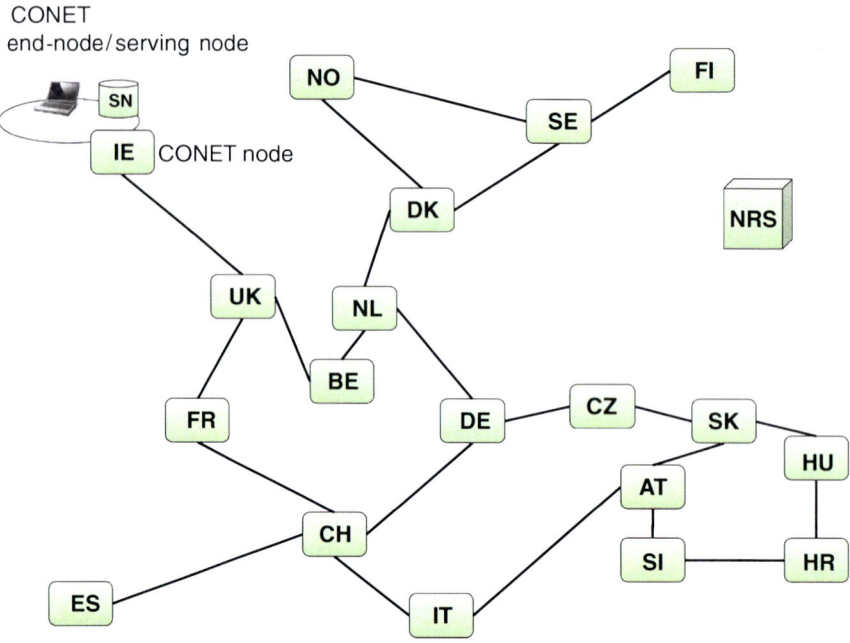

Fig. 3.13 PlanetLab CONET overlay deployment

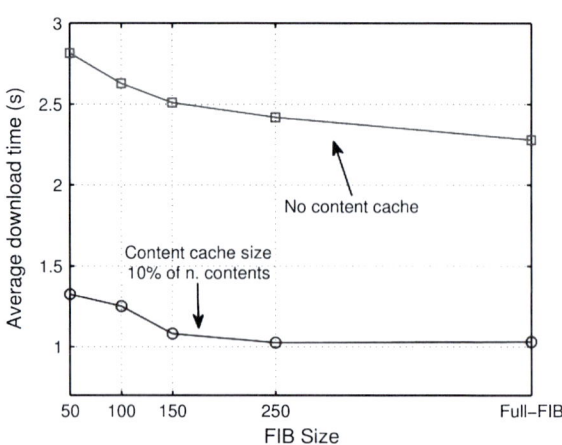

Fig. 3.14 Average download time versus node FIB size in laboratory test-bed, with and without in-network caching

processing delay brought about by the NRS node, which in the worst case is equal to about 350 ms. This considerable delay penalty was not present in the laboratory test-bed of Fig. 3.12, since there we had a direct Ethernet connection between the node and the NRS. Conversely, in the PlanetLab test-bed, NRS is connected to nodes through a best-effort Internet path, which introduces a significant delay.

We argue that this delay is due to the connectivity/processing delay brought about by the NRS node. Such delay (in the worst case equal to about 350 ms) would not occur if the traffic from/to the NRS had priority on the other user traffic and if the NRS were implemented by using a suitable powerful hardware.

3.10.2 Transport of Named-Data

As discussed in Sect. 3.6.1, the CONET adopts a receiver-driven protocol where endpoints exchange Interest/Named-Data CPs sequences and the exchange rate is regulated by the receiver, following principles proposed in (Salsano et al. 2012). This protocol, which we generically name Information-Centric Transport Protocol (ICTP), implements the same algorithms of TCP (slow-start, congestion avoidance, fast retransmit, fast recovery), but adapted to the receiver-driven operation.

In (Salsano et al. 2012), we measured the performance improvement provided by CONET ICTP with respect to the transport mechanism included in the CCNx tool.

In short, the main differences between our ICTP and the transport protocol provided by the CCNx tool can be summarized as follows:

(1) *Use of carrier-packets*: we can say that the data-units of the CCNx transport protocol correspond to the named-data CIUs (aka Data messages), each of which is relatively large, (e.g. 4 kB). This means they have to be segmented at the IP level. The use of such a large data-unit decreases the efficiency and promptness of congestion control. To overcome this problem, ICTP divides named-data CIUs into segments, each of which is transported by a carrier-packet, whose size is close to that of a TCP segment. In other words, the carrier-packet is the ICTP data-unit and no longer the whole named-data CIU. This suggests that the performance of the transport layer will be similar to that of the transport layer in TCP.

(2) *Mimicking TCP congestion control*: CNNx implements a simple congestion control algorithm that resembles the traditional Selective-Repeat ARQ with fixed window size. Conversely, ICTP thoroughly mimics the TCP Reno congestion control algorithms.

We evaluate the performance of ICTP in a laboratory scenario with a single sub-system; in the scenario an end-node fetches named-data from a serving node with a 10 Mb/s link. We compared the ICTP performances with the performances of the congestion control in CCNx. Figure 3.15 shows the application goodput measured for CCNx and for ICTP versus the IP packet loss probability and for two different sizes of the named-data CIUs (4 kB and 32 KB). In the ideal lossless case, the performance of the current CCNx transport implementation is slightly better, as it has a smaller overhead (not using carrier-packets). The performance of ICTP becomes better as soon as packet loss is introduced.

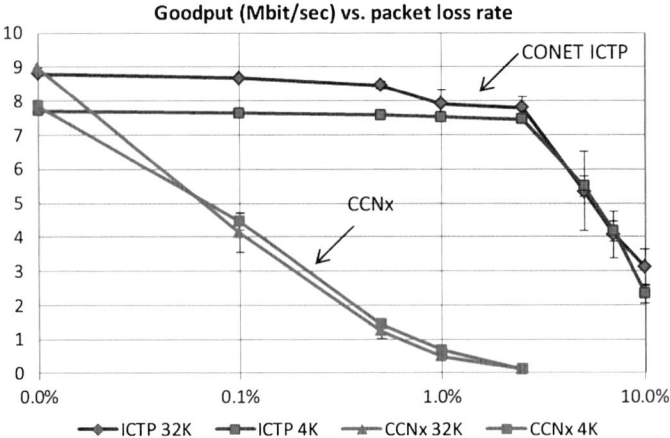

Fig. 3.15 CONET versus CCNx transport of named-data

3.11 Conclusions

Our aim in designing the CONET was to achieve the promised advantages of ICN (listed in Chap. 1), while mitigating its two main drawbacks: (1) ICN requires changes in the basic network operation, which per se is already a big obstacle to take-up of this approach; (2) ICN raises scalability concerns, as the number of different contents and corresponding names is much bigger than the number of host addresses; this has obvious implications on the size of routing tables and on the complexity of lookup functions; in addition, in some proposed CCN architectures (Jacobson et al. 2009b), providing reverse paths (for information delivery) requires maintaining states in network nodes.

As for the first point, we looked for graceful incremental solutions, backward compatible with the current Internet, as opposed to risky clean slate and flag-day solutions. As regards the second point, we proposed some specific solutions mitigating this problem.

More in details: our routing-by-name architecture is able to support current and future Internet traffic with off-the-shelf technologies; our transport mechanism has performance close to those of the current TCP; our API is similar to that of HTTP facilitating the development of HTTP-to-CONET proxy for the support of legacy applications; our push-based services allow supporting legacy interactive services. Thus, we believe that our proposed CONET can allow an advantageous and smooth migration from the current Internet based on the TCP/IP API, towards a future Internet based on the ICN API.

References

A. Detti, N. Blefari Melazzi, S. Salsano, M. Pomposini, "CONET: A Content Centric Inter-Networking Architecture", ACM SIGCOMM Workshop on Information-Centric Networking (ICN 2011), August 19, 2011a, Toronto, Canada.

A. Detti, S. Salsano, N. BlefariMelazzi, "IPv4 and IPv6 Options to support Information Centric Networking", Internet Draft, draft-detti-conet-ip-option-02, Work in progress, October 2011b.

A. Detti, M. Pomposini, N. Blefari Melazzi, S. Salsano, "Supporting the Web with an Information Centric Network that Routes by Name", Elsevier Computer Networks, vol. 56, issue 17, 2012a, p. 3705–3722.

A. Detti, A. Caponi, N. Blefari-Melazzi," Exploitation of Information Centric Networking Principles in Satellite Networks", IEEE ESTEL 2012, Roma, Italy, 2-5 October 2012b.

A. Detti, M. Pomposini, N. Blefari Melazzi, S. Salsano, A. Bragagnini, "Offloading cellular networks with Information-Centric Networking: the case of video streaming", IEEE International Symposium on a World of Wireless, Mobile and Multimedia Networks 2012c.

A. Detti, A. Caponi, G. Tropea, N. Blefari-Melazzi, G. Bianchi, "On the Interplay among Naming, Content Integrity and Caching in Information Centric Networks", submitted for publication 2013, available at. http://netgroup.uniroma2.it/Andrea_Detti/papers/conferences/ICN-Naming-Signature-Caching-Interplay.pdf.

A. Ghodsi, T. Koponen, B. Raghavan, S. Shenker, A. Singla, and J. Wilcox, "Information-Centric Networking: Seeing the Forest for the Trees", in Proc. of the 10th ACM Workshop on Hot Topics in Networks (HotNets-X), November 14-15, 2011, Cambridge, MACambridge, Massachusetts.

A. Kuzmanovic, E.W. Knightly. "Receiver-Centric Congestion Control with a Misbehaving Receiver: Vulnerabilities and End-point Solutions", Elsevier Computer Networks. 2007, 51, 2717–2737.

CCNx project web site: http://www.ccnx.org.

D. Cheriton, M. Gritter, "TRIAD: a scalable deployable NAT-based internet architecture", Technical Report (2000).

D. Perino, M. Varvello, "A Reality Check for Content Centric Networking", ACM SIGCOMM Workshop on Information-Centric Networking (ICN 2011), August 19, 2011, Toronto, Canada.

D. Smetters, V. Jacobson: "Securing Network Content", PARC technical report, October 2009.

D. Trossen, M. Sarela, and K. Sollins: "Arguments for an information-centric internetworking architecture" SIGCOMM Computer Communication Review, vol. 40, pp. 26-33, 2010.

K. Changhoon, A. Caesar, A. Gerber, and J. Rexford. "Revisiting route caching: The world should be flat." Passive and Active Network Measurement (2009): 3-12.

L. Chiariglione, A. Difino, N. Blefari Melazzi, S. Salsano, G. Tropea, A. C. G. Anadiotis, A. S. Mousas, I. S. Venieris, C. Z. Patrikakis: "Publish/Subscribe over Information Centric Networks: a Standardized Approach in CONVERGENCE", Future Network & Mobile Summit 2012, 4 - 6 July 2012, Berlin, Germany.

L. Galluccio, G. Morabito, S. Palazzo, "Caching in information-centric satellite networks", IEEE ICC 2012, June 2012, Ottawa, Canada.

L. Popa, A. Ghodsi, and I. Stoica. 2010 "HTTP as the narrow waist of the future internet", in Proceedings of the 9th ACM SIGCOMM Workshop on Hot Topics in Networks (Hotnets-IX). ACM, New York, NY, USA.

N. Blefari-Melazzi, A. Detti, G. Morabito, S. Salsano, L. Veltri," Supporting Information-Centric Functionality in Software Defined Networks", IEEE International Conference on Communications (ICC 2012a).

N. Blefari Melazzi, A. Detti, M. Pomposini, S. Salsano: "Route discovery and caching: a way to improve the scalability of Information-Centric Networking", IEEE Global Communications Conference 2012, Anaheim, California, Dec., 3-7 2012b.

N. Blefari Melazzi, A. Detti, M. Pomposini, "Scalability Measurements in an Information-Centric Network", Measurement-based experimental research: methodology, experiments and tools", Springer Lecture Notes in Computer Science (LNCS), vol. 7586, 2012c.

N. McKeown, T. Anderson, H. Balakrishnan, G. Parulkar, L. Peterson, J. Rexford, S. Shenker, and J. Turner, "OpenFlow: Enabling Innovation in Campus Networks". White paper. March 2008 (available at: http://www.openflow.org).

Patrick Th. Eugster, Pascal A. Felber, Rachid Guerraoui, and Anne-Marie Kermarrec. "The many faces of publish/subscribe", ACM Comput. Surv. 35, 2 (June 2003), 114-131.

S. Salsano, A. Detti, M. Cancellieri, M. Pomposini, N. Blefari Melazzi, "Transport-layer issues in Information Centric Networks", ACM SIGCOMM Workshop on Information-Centric Networking (ICN 2012), August 17, 2012, Helsinki, Finland.

T. Koponen, M. Chawla, B.G. Chun, A. Ermolinskiy, Kye Hyun Kim, S. Shenker, I. Stoica: "A data-oriented (and beyond) network architecture", Proc. of ACM SIGCOMM 2007, August 27-31, Kyoto, Japan.

V. Jacobson, et al. "Networking named content", in Proc. of ACM CoNEXT 2009a, December 1-4. Rome, Italy.

V. Jacobson, et al. "VoCCN: voice over content-centric networks", Proceedings of the 2009 Workshop on Re-architecting the Internet (ReArch 2009); 2009 December 1-4; Rome, Italy. NY: ACM; 2009b.

X. Zhao, D. J. Pacella, and J. Schiller, "Routing Scalability: An Operator's View", IEEE Journal on Selected Areas in communications, vol. 28, no. 8, October 2010.

Chapter 4
The Content Level (CoMid)

Angelos-Christos G. Anadiotis, Aziz S. Mousas, Angelo Difino and Charalampos Z. Patrikakis

Abstract This chapter provides the description of the overall architecture of the content level of convergence: the CONVERGENCE Middleware (CoMid). The chapter starts with the presentation of the MPEG-M standard, which provides the foundations for CoMid, then proceeding with a complete description of this architectural level. The key components of the Content level comprise a diversified set of middleware engines to manipulate Versatile Digital Items (VDIs), a Community Dictionary Service (CDS) and a Semantic Overlay. The set of middleware engines were partially borrowed and/or adapted from MPEG-M and partially designed and developed from scratch within CONVERGENCE, adopting the same design principles. The CDS and the Semantic Overlay, newly designed and developed by CONVERGENCE, when used together with the middleware engines, enable the semantic, content-based, publish-subscribe functionality of the platform.

4.1 Introduction

Whereas the CONVERGENCE network level handles named-resources (named-data or named-service-access points) only in terms of their names, the CONVERGENCE content level offers functionality to handle resources on the basis of their content or offered services. These functionalities are implemented using of a set of technologies that we call CONVERGENCE Middleware (CoMid), which comprises a set of middleware engines, the Community Dictionary Service (CDS) and a Semantic Overlay.

A.-C. G. Anadiotis (✉) · A. S. Mousas · C. Z. Patrikakis
National Technical University of Athens, Athens, Greece
e-mail: aca@icbnet.ece.ntua.gr

A. Difino
CEDEO.net, Turin, Italy

F. Almeida et al. (eds.), *Enhancing the Internet with the CONVERGENCE System,*
Signals and Communication Technology, DOI: 10.1007/978-1-4471-5373-3_4,
© Springer-Verlag London 2014

To understand the content of named-data and the service offered by a named-service-access-point, the CONVERGENCE system describes data and service-access-points using metadata in a structure called Versatile Digital Item (VDI). The content level provides the necessary functionality for the management of VDIs, i.e. to create/edit, publish and un-publish VDIs. More information on the VDIs can be found in Chap. 5.

The separation between the Content Level and Network Level is based on the conceptual separation between metadata and data. If metadata needs to be moved around, stored, searched, interpreted or updated, this is a task for the CoMid. When data has to be stored, retrieved, replaced or distributed, this is a task for the CoNet, which is the CONVERGENCE component of the network level (see Chaps. 2 and 3 for more details on the CONVERGENCE architecture and the CONVERGENCE network level). This distinction clearly separates functionality that needs to "understand" VDIs (implemented in the CoMid) from functionality that only needs to know the name of the resources it is handling (implemented in the CoNet). Some VDIs may only contain metadata describing given data (e.g. a picture of the Louvre museum), and information on how to retrieve the data (e.g. a Network Identification, NID). Others may also contain the data itself. In this latter case, the resource (e.g. a picture) is embedded within the VDI.

The CDS can be regarded as the CONVERGENCE knowledge base. The CDS contains the ontologies used to perform semantic search. When the user submits subscriptions to CONVERGENCE, indicating satisfying certain search criteria or preferences, the CDS provides CoMid with all matching concepts. CoMid can then perform matching with available VDIs.

The Content Level supports content-based resource-discovery. A user expresses interest in certain kinds of resources by specifying a set of criteria for resource attributes. The resource-discovery functionality then returns a list of VDIs matching the search criteria. Some resource attributes may contain semantic information. By interacting with the CDS infrastructure, the content level provides semantic and content-based resource-discovery. The middleware engines of CoMid are able to implement semantic-based resource-discovery by analyzing the metadata of available VDIs distributed inside the CoMid, using the matching concepts provided by the CDS.

A user can perform (semantic) content-based resource-discovery either by making a search request, or through a subscription. In the case of a search request, the system serves the request and forgets everything about it. In the case of a subscription, the content level provides a content-based subscription that makes the discovery instance permanent. Every time the system finds a new resource fitting subscriber search criteria, it delivers a notification to the subscriber.

In addition to resource-discovery and subscription functionality, the Content Level also includes VDI management services (for example, creating, packaging, storing, rights management, and delivery). This functionality is either directly derived from the MPEG-M framework (MPEG eXtensible Middleware), or offered as dedicated building blocks of CONVERGENCE, compatible with MPEG-M.

The CoMid functionality and architecture, detailed in the following sections, can thus be considered as a CONVERGENCE-specific part of a wider, standard, MPEG framework of services.

4.2 CONVERGENCE Middleware

The CoMid architecture is based on concepts developed in the MPEG-M standard, an extension of the former MXM MPEG platform proposed by MPEG. CoMid functionality is implemented using Elementary Services in MPEG-M (Fig. 4.1).

CoMid is based on a DDD (Describe, Discover and Distribute) paradigm. Its main purpose is to offer a set of standardized building blocks, APIs and protocols to describe, discover and distribute resources. Higher-level tools and applications are built by assembling CoMid blocks into custom software, manipulated by users. These tools lie at the Application level of the CONVERGENCE architecture (see Chap. 2 for a description of the CONVERGENCE architecture and Chap. 8 for details on the CONVERGENCE Applications).

4.2.1 The MPEG-M Standard

Before continuing with the description of CoMid and its operation, it is important to provide an overview of MPEG-M, which is the standard on which CoMiD is based. The emerging MPEG-M specification (ISO/IEC 23006—Multimedia Service Platform Technologies) is a suite of standards providing Architecture, Technology

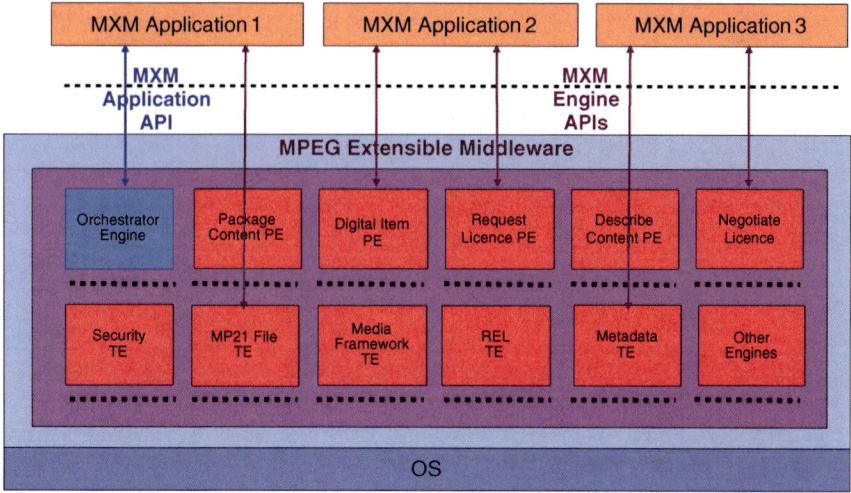

Fig. 4.1 The MXM standard

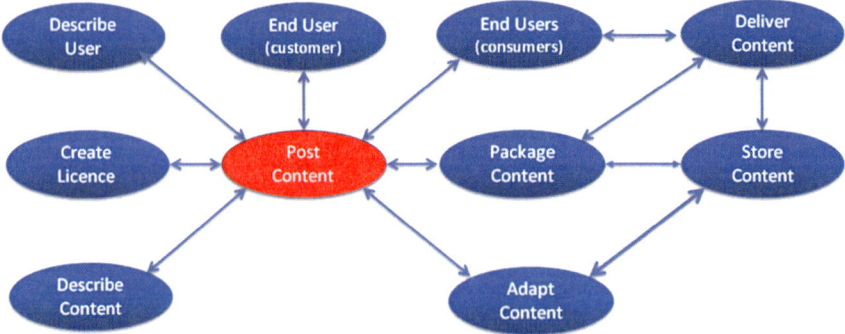

Fig. 4.2 A possible service chain centred on post content SP

Engines (TE), Protocol Engines (PE), Aggregation Technologies and Reference Software that are well aligned to the basic concepts of CONVERGENCE (which were presented in Chap. 2, specifically illustrated in Fig. 4.2 of that chapter, "Overview of CONVERGENCE architectural levels"). The standard includes:

- Part 1 Architecture
- Part 2 MPEG Extensible Middleware (MXM) API
- Part 3 Conformance and reference software
- Part 4 Elementary services
- Part 5 Service aggregation.

The figures representing the peer architecture depicted in this chapter are based on MPEG-M. Part 2, referred as MXM, introduces the notion of Engine, i.e. appropriate groupings of technologies and corresponding Application Programming Interfaces (API) through which an Application can access the functionality it needs. As an Application typically needs more than one engine ("chains" of engines), MXM also provides examples of Orchestrator Engines, special MXM Engines capable of creating chains of Engines to execute, high-level application calls such as "Play" (see Fig. 4.1).

Part 4 introduced the notion of Elementary Services and specifies the messages exchanged between two peers. An implementation of protocols as specified by Part 4 is called a Protocol Engine. Finally Part 5 specifies how Protocol Engines can be chained to provide so-called Aggregated Services.

In recent years, many new digital media related services have appeared. Many such services are actually combinations of Elementary Services. MPEG has seen that, standardizing a set of technology elements and communication protocols facilitates the creation of aggregated services, from a set of standard Elementary Services, even on demand.

Assuming that in there is a Service Provider (SP) for each Elementary Service, a User may ask the Post Content SP to get a sequence of songs satisfying certain Content and User Descriptions. The Fig. 4.2 below depicts how a chain of services can respond to the user's request.

Then End User contacts Post Content SP, which gets appropriate information from Describe Content SP and Describe User SP to prepare the sequence of songs using its internal logic. The End User subsequently gets the necessary licenses from Create License SP. The sequence ("titles") of songs is handed over to Package Content SP. Package Content SP gets the songs ("Resources") from Store Content SP and hands over the Packaged Content to Deliver Content SP which streams the Packaged Content to End User.

MPEG has specified a set of standard Elementary Services and related protocols to enable distributed applications to exchange information about entities playing a role in digital media services (e.g. Content, Contract, Device, Event and User), and the processing that a party may wish to execute on those entities (e.g. Authenticate, Create, Deliver, Describe, Identify, Negotiate, Process, Request, Search and Transact). These have been standardized in part 4 "Elementary Services" (ISO 23006-4).

Based on these advantages, CONVERGENCE has decided to adopt MPEG-M as the basis of its middleware architecture, including MPEG-M protocol and technology engines, as well as aggregation technologies. Many CONVERGENCE Protocol and Technology Engines come in fact from MPEG-M.

Additionally, given the necessity of supporting all of the requirements identified for CONVERGENCE, the project has developed new engines and extended the functionality of some of the existing MPEG-M engines. In particular, it has extended the Describe Content PE to support integration with ontologies (CDS) and the Security engine to offer the levels of security envisaged by CONVERGENCE.

Accordingly, CONVERGENCE has designed and developed new Protocol and Technology engines to support additional functionality, notably:

(a) Protocol Engines

1. Describe Content (extended functionality)
2. Post Content (new)—included in the standard

(b) Technology Engines

1. CDS—included in the standard
2. CoNet
3. Match—standardized in the context of MPEG-M Search TE
4. Overlay—included in the standard
5. Security (extended functionality).

4.2.2 Overview of Key CoMid Features

The following sections of the chapter briefly introduce key CoMid functionalities, making it possible to:

- Operate on VDIs, describing resources with metadata extracted from well-known or custom semantic taxonomies of concepts;
- Publish information on resources into an overlay of peers whose topology is based on those same semantic taxonomies; this overlay will be referred to as Semantic Overlay;
- Search for resources in specified regions of the Semantic Overlay and deliver matching content to users, even when resource descriptions do not fully match user requests.

The VDI

CONVERGENCE users act on VDIs. These are XML structures containing:

1. Identifiers
2. (Links to) Resources
3. (Links to) Semantically-rich metadata describing resources
4. (Links to) Licenses expressing which rights are given to act on resources
5. Event Report Requests (ERR) instructing a peer to issue an Event Report (ER) when a particular event (e.g. play, store or match) occurs.

CONVERGENCE has defined four types of VDI: Resource (R–VDI), Publication (P–VDI), Subscription (S–VDI) and User (U–VDI). As part of the creation process, a VDI is assigned a unique and persistent identifier. When a VDI is "superseded" by a new version or when a new VDI is related to an existing one, the CONVERGENCE Ontology Services let users establish links between them.

Semantic and Dictionaries

The CONVERGENCE Core Ontology (CCO) is a native CONVERGENCE component. The CCO allows for a semantic organization of peers in a virtual overlay network of "fractals", which are dynamically shaped and connected, on the basis of users' interests in different types of content. Peers join or leave a fractal based on what users are currently publishing and subscribing to. The basic structure of the CONVERGENCE Semantic Overlay is shown in Fig. 4.3. A peer typically belongs to more than one fractal, depending on how many users are interested in what kind of content. To provide redundancy and mitigate peer churning (peers may be offline when users wish to access them), fractals are usually populated by more than one peer.

Domain and user ontologies can also be used to define fractals. Typically the former come from Service Providers (SP), whereas the latter are created by individual users. Therefore, the fractal organization can be seen as covering multiple "dimensions", where each dimension is defined by an ontology.

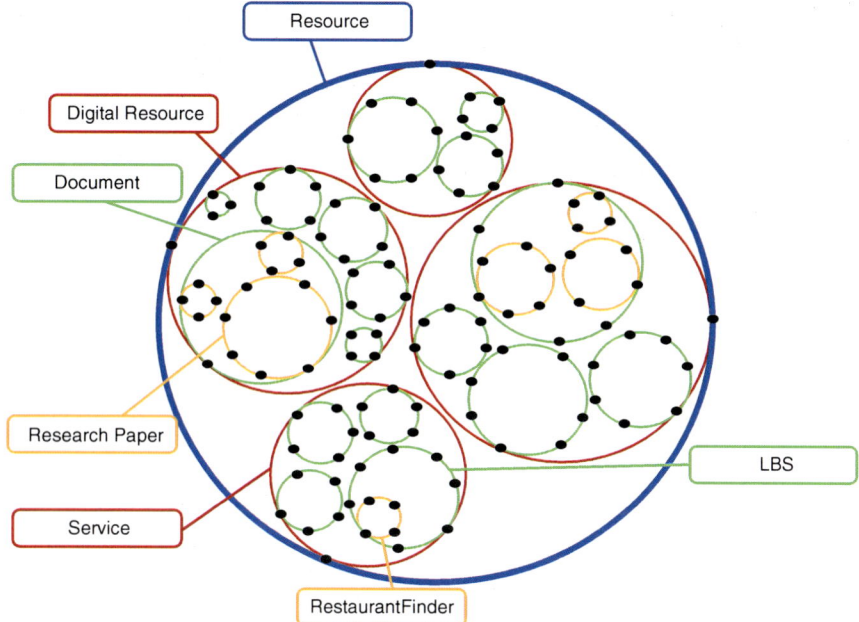

Fig. 4.3 An example of structure of CONVERGENCE fractals

Thus, a peer may reference resources that were described using concepts based on different ontologies, in different dimensions.

The CDS component is also part of the middleware and is implemented as a Technology Engine (CDS TE). This component maintains *dictionaries* that help translate concepts and properties from one ontology model to another. The CDS is exploited when users describe resources and when descriptions of what is being published do not match the terms used in user definitions of search criteria. In this cases, the CDS attempts to translate betweens concepts and resolve the match.

Publish/Subscribe

The key interaction pattern for CONVERGENCE users is based on a publish/subscribe paradigm. Users who make resources available to the system and discoverable are considered to publish them. Users searching for specific resources are considered to subscribe to them. The interaction is asynchronous and decoupled in time and space. Publications and subscriptions are both described by semantically-rich metadata and labelled with relevant concepts. To find relevant content, a dedicated CoMid engine executes a semantic query, representing the subscription. The query is successful when published content matches the subscription criteria.

Fractals in CONVERGENCE Semantic Overlay represent focused concepts, i.e. unions and/or intersection of concepts in the core ontology or other ontologies (dimensions). All P-VDIs and S-VDIs, carrying publications and subscriptions labelled with a certain concept, are stored in a specific set of peers in the fractal. Spatial redundancy (e.g. using more than one peer to store information) is used to ensure propagation of information, because at any given time some peers might be inactive. The use of "focused" fractals makes it possible to restrict the search space to peers actively participating in the fractal.

Peers have the embedded ability to:

1. Perform matches between P-VDIs and S-VDIs
2. Communicate any match to specified peers in the form of ER (Event Reports), depending on licenses and ERRs (Event Report Requests)
3. Remove S-VDIs and P-VDIs from the match tables, if one of the two following situations occurs:
   - their expiration date has passed;
   - an authorized user requests to remove them before the expiration date.
4. Aggregate ERs from different peers and communicate the result to the end user.

4.2.3 Community Dictionary Service

The CDS is the CoMid component responsible for supporting CONVERGENCE semantic functionalities. To support this role, the CDS maintains and exploits ontologies and the relationships between them. CDS is a distributed service: each peer has access to a local CDS that might be customized with user ontologies, and/ or can use more complete CDSs to be accessed remotely.

The following sections briefly describe how the CDS supports semantic descriptions of VDIs and the way these descriptions are exploited to provide semantic matching between publications and subscriptions.

CDS Support for Content Description

Given Convergence's content-centric publish/subscribe paradigm, it is vital to correctly match published VDIs to user subscriptions. This requires coherent semantic descriptions of VDI resources and users subscription criteria. Users employ CDS servers that contain ontology models that can accurately describe the resource they wish to publish and to subscribe to. User can run their own (local or remote) CDS servers, each containing custom/users ontologies as well as well-known domain ontologies. The following paragraphs contain details on the deployment and use of the CDS.

CDS Involvement in Publishing

To help users in building semantic descriptions of their VDIs, the CDS exposes a service for ontology entity exploration. This service allows users to select ontology entities to semantically describe their VDIs. The walkthrough below describes the creation of the semantic description of a VDI for the movie "Star Wars". The user's CDS is loaded with the Movie Ontology[1] and the IMDB ontology (Grimnes 2012). The identifier of the VDI is set to value RVDI_23 (Table 4.1).

CDS Involvement in Subscribing

The same service can help users to define subscription criteria. The walkthrough below presents the creation of a SPARQL query for the movie "Star Wars". User's CDS is again loaded with the Movie Ontology and the IMDB ontology (Table 4.2).

CDS Dictionaries

Matching is one of the core challenges that the CONVERGENCE system has to address. The plethora of ontology models and the freedom granted to users to select the ontology with which they describe their VDIs (or to use their own selection of ontologies), inevitably creates a huge diversity in the metadata used to described VDIs and the risk that user publications and subscriptions may not match unless they are both described using the same ontology.

CDS dictionaries are a new concept designed to address the issues of heterogeneous metadata. In essence, a CDS dictionary is an ontology just like user ontologies and domain ontologies. However, unlike other ontologies, its content consists of a mapping between two different ontologies: a set of equivalence statements between entities in the first ontology and entities in the second. In this way, CDS dictionaries provide semantic bridges between user-ontologies, between user-ontologies and domain ontologies or between domain ontologies.

Figure 4.4 depicts the set of ontologies the CDS uses during publication, subscription and matching procedures.

CDS Dictionary Format

As an example of the CDS dictionary format, the table below provides a mapping between the movieOntology and the IMDB ontologies (Table 4.3).

---

[1] www.movieontology.org.

Table 4.1 Content publication workflow with CDS

User action	System response	Semantic description
User types "movie"	CDS searches for an entity named movie and returns the IMDB:Movie and the movieOntology:Movie Classes	N/A

User selects movieOntology:Movie	System selects the movieOntology ontology for the description of the VDI and returns the RDF description	`<rdf:rdf>`
		`<rdf:description rdf:about="RVDI_23">`
		`<rdf:type rdf:resource="&movieOntology;Movie"/>`
		`</rdf:description>`
		`</rdf:rdf>`

User types title	CDS returns with the movieOntology:title DatatypeProperty	N/A
User selects movieOntology:title	System prompts user to enter the title of the movie	N/A

User types "Star Wars"	System returns to the user the RDF description	`<rdf:rdf>`
		`<rdf:description rdf:about="RVDI_23">`
		`<rdf:type rdf:resource="&movieOntology;Movie"/>`
		`<movieOntology:title>`
		Star Wars
		`</movieOntology:title>`
		`</rdf:description>`
		`</rdf:rdf>`

Table 4.2 Content subscription workflow with CDS

User action	System response	Semantic query
User types movie	CDS searches for an entity named movie and returns the IMDB:Movie and the movieOntology:Movie Classes	
User selects IMDB:Movie	System selects the IMDB ontology for the description of the VDI and returns the sparql query	SELECT ?x WHERE { ?x rdf:type IMDB:Movie }
User types title	CDS returns with the IMDB:title DatatypeProperty	
User selects IMDB:title	System prompts user to enter the title of the movie	
User types "Star Wars"	System returns the user the SPARQL query	SELECT ?x WHERE { ?x rdf:type IMDB:Movie ?x IMDB:title "Star Wars" }

Fig. 4.4 CDS ontologies

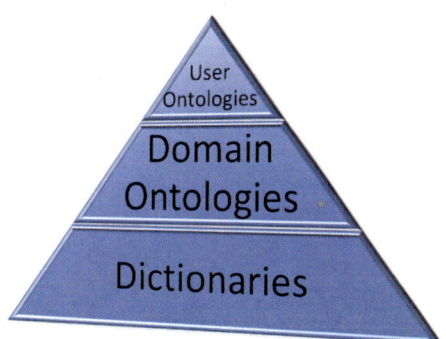

Dictionaries can also use the skos:narrower and skos:broader properties to connect ontology entities, in case they are not semantically equivalent.

Mechanisms for Exploiting CDS Dictionaries

Publishing As mentioned earlier, in the particular example for the explanation of the use of CDS, we are considering the use of the "Star Wars" Resource VDI. The metadata in the VDI is based on the movieOntology. The walkthrough below shows how the user can use the CDS to create semantically equivalent metadata based on the IMDB ontology (Table 4.4).

Table 4.3 Ontology mapping example

Ontology	Mapping
MovieOntology ontology	<owl:Class rdf:about="&movieOntology;Movie"/> <owl:DatatypeProperty rdf:about="&movieontology;title"> <rdfs:domain rdf:resource="&movieOntology;Movie"/> <rdfs:range rdf:resource="&xsd;string"/> </owl:DatatypeProperty> ...
MovieOntology-IMDB dictionary	<owl:Class rdf:about="&movieOntology;Movie"> <owl:equivalentClass rdf:resource="&IMDB;Movie"/> </owl:Class> <owl:DatatypeProperty rdf:about="&movieontology;title"> <owl:equivalentProperty rdf:resource="&IMDB;title"/> </owl:DatatypeProperty> ...
IMDB ontology	<owl:Class rdf:about="&IMDB;Movie"/> <owl:DatatypeProperty rdf:about="&IMDB;title"> <rdfs:domain rdf:resource="&IMDB;Movie"/> <rdfs:range rdf:resource="&xsd;string"/> </owl:DatatypeProperty> ...

Table 4.4 CDS mapping example in publishing

User action	System response	Semantic description
User passes the movieOntology-based semantic description to the CDS		<rdf:rdf> <rdf:description rdf:about="VDI1"> <rdf:type rdf:resource="&movieOntology;Movie"/> <movieOntology:title > Star Wars </movieOntology:title> </rdf:description> </rdf:rdf>
	CDS returns the user the IMDB-based equivalent RDF description	<rdf:rdf> <rdf:description rdf:about="VDI1"> <rdf:type rdf:resource="&IMDB;Movie"/> <IMDB:title > Star Wars </IMDB:title> </rdf:description> </rdf:rdf>

Subscribing The CDS can also be used to translate a SPARQL query from one ontology to another, as shown in the walkthrough (Table 4.5).

Matching Upon receiving the publication, the CDS is requested to expand the semantic description of the VDI of the movie using the movieOntology-IMDB dictionary. As seen in the walkthrough below, this procedure creates new triples (Table 4.6).

After expansion, movieOntology-based and the IMDB-based SPARQL queries will both find the match. This mechanism is explained in detail in the Sect. 4.4.

Table 4.5 CDS mapping example in subscribing

User action	System response	Semantic query

User passes the IMDB-based SPARQL query to the CDS		SELECT ?x
		 WHERE
		 {
		 ?x rdf:type IMDB:Movie
		 ?x IMDB:title "Star Wars"
		 }

	CDS returns the user the movieOntology-based equivalent SPARQL query	SELECT ?x WHERE { ?x rdf:type movieOntology:Movie ?x movieOntology:title "Star Wars"}

Table 4.6 CDS mapping example in matching

User action	System response	Semantic description
User passes the movieOntology-based semantic description to the CDS for materialization		\<rdf:rdf\> \<rdf:description rdf:about="VDI1"\> \</rdf:description\> \</rdf:rdf\>
	CDS returns the user the implemented RDF description of the VDI	\<rdf:rdf\> \<rdf:description rdf:about="VDI1"\> \<rdf:type rdf:resource="&IMDB;Movie"/\> \<IMDB:title \> Star Wars \</IMDB:title\> \<rdf:type rdf:resource="&movieOntology;Movie"/\> \<movieOntology:title \> Star Wars \</ movieOntology:title\> \</rdf:description\> \</rdf:rdf\>

4.2.4 Semantic Overlay

Semantic Foundations

CONVERGENCE is built upon a content-centric network, which is accessed through the CoNet engine and the corresponding APIs. Hence, applications are not aware of the locations where resources are stored. To access a resource, they use the name under which it is advertised in CoNet. However, a name cannot convey the whole meaning of the resource, which is necessary for discovering the resource using some (semantic) description.

The indexing and efficient retrieval of content by semantic metadata is the key feature of the CONVERGENCE **Semantic Overlay**.

Efficient semantic operations on metadata require a scheme for semantic categorization of resources (Muhr et al. 2010). The first element in the CONVERGENCE

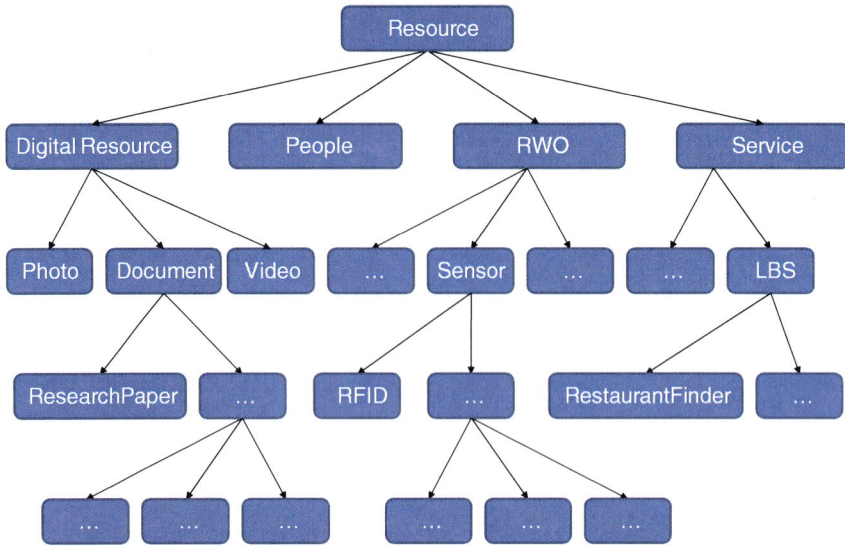

Fig. 4.5 CONVERGENCE core ontology (CCO)

scheme is the CONVERGENCE Core Ontology (CCO): a CONVERGENCE-wide hierarchical taxonomy of resource semantic types, implemented as an OWL ontology. The first level root concept in the ontology is the "Resource", while first level children are "Digital Resources", "People", "Real World Objects" and "Services", corresponding respectively to the *Internet of Media,* the *Internet of People,* the *Internet of Things and* the *Internet of Services.* Figure 4.5 provides an abstract view of the CCO.

The CCO is not the only ontology available to CONVERGENCE users. Additional ontologies make it possible to categorize content in other ways, which may either make it more specific (e.g. it could be an ontology of music genres or movie types) or they could refer to entirely different context (e.g. classifying the peers based on their location).

Semantically Managing the Overlay Topology

Peers belonging to the overlay are partitioned into groups based on semantic criteria. For example, all peers injecting VDIs which belong to the category cco:movie, belong to the group cco:movie. Groups like this have the same kind of self-similarity we find in fractals.[2] We therefore call them *network fractals.*

---

[2] Fractals: Geometric shapes "that can be split into parts, each of which is (at least approximately) a reduced-size copy of the whole" (Mandelbrot 1982).

A peer may belong to some fractals in the CCO and other fractals in another ontology, representing another *dimension* of semantic meaning (e.g. an ontology of access rights). This organization is *virtual* and not a physical one. Each peer propagates a message in one or more fractals, in one or more dimensions. Every time a peer decides to propagate a message to the overlay, it determines the final group of peers that should receive the message. These peers may belong to one or more fractals in one or more dimensions.

Consider a fractal i belonging to the j-th dimension and consisting of a group of peers, G_i^j. The message is then propagated to the peers of the group G that satisfy the condition:

$$G = \bigcap^j \bigcup_i G_i^j$$

This implies that:

- When a user wants to insert content into a set of fractals belonging to a single dimension, he/she sends this content to the peers of the union of these fractals.
- When a user wants to insert content into a set of fractals belonging to multiple dimensions, she takes the union of fractals for each dimension and injects the content into the intersection of these sets.

Propagation Protocol

A major challenge facing CONVERGENCE Semantic Overlay is the dynamic nature of the CONVERGENCE system, quite similar to that faced by P2P systems. Peers may continuously join and leave fractals for many different reasons:

- A peer is no longer interested in a topic (i.e. no publications or subscriptions for that topic are present in the peer) and leaves the fractal.
- A new peer is interested in a topic (makes a publication and/or subscription for that topic) and enters the fractal.
- A peer goes offline/online.
- A peer crashes.

For this reason, CONVERGENCE employs a gossiping protocol in order to propagate content in the fractal and to enable peers to maintain a partial view of their fractals. Since gossiping protocols have been proven to be robust and scalable, they provide a good solution for dynamic networks (Eugster et al. 2004; Friedman et al. 2007; Ganesh et al. 2001; Kermarrec et al. 2003).

Peers Registration to Fractals

Each fractal maintains a registry, which is maintained and used by constituting peers belonging to the fractal to communicate with each other. The registry contains the following important information.

1. **The URI of the ontology** used to partition the system into fractals. This information is used to access fractals that are higher or lower in the hierarchy than the current fractal.
2. **The fractal identifier**. This shows which fractal the registry refers to.
3. **The size of the fractal** (the number of peers in the fractal). This number is used to determine the size of this registry.
4. Other characteristics of peers belonging to the fractal

   - **The peer identifier.**
   - **The peer overlay propagation service endpoint**. Publication and subscription VDIs are pushed to this peer via this service endpoint.
   - **The date on which the peer is scheduled to leave the fractal** (the latest expiration date for any publication or subscription a peer has injected into the fractal).

Peers remain in the registry until the leave date has passed. When a peer receives a VDI from a remote peer, it checks the VDI's expiry date and updates the leave date of the remote peer in the local registry accordingly. After a peer has selected the peers to which it will propagate content (*partial view*), it checks their leave dates. In case there are peers in the partial view whose leave dates have passed, the peer removes them from the partial view and the registry. Finally, it replaces the removed peers in the partial view with others from the registry.

Each peer periodically advertises a part of the registry under a given, standard name, which is decided a priori. The selection of the name is system-dependent; for example, in CONVERGENCE, the registry for a fractal of type cco:movie, has the name urn:overlay:registry:cco:movie. This way, any peer that enters the system will always find another peer that can provide it with an entry point to the system. Since gossiping protocols do not require from a peer to communicate with all the other peers of the fractal, rather than with $O(logN)$ of them, where N is the size of the fractal, the size of the advertised registry is of order $O(logN)$.

Next, we are describing how the topology management protocol of the overlay handles the creation of a fractal and the update of the registry:

- **Bootstrapping.** When a peer enters a fractal (by issuing a publication or a subscription), it first has to register. To do this, it first requests the fractal registry by its name (as mentioned above, each peer periodically advertises a part of its registry). If there is no registry, this means that it is the first peer in the fractal. It therefore creates a registry, adds itself to the registry and advertises the registry to the network. If a registry already exists, it gossips a discovery message to the peers in the registry, which continue the gossiping of the discovery message. Each peer receiving a discovery message immediately adds the initiator (the first peer in the path contained in the header) to its registry, increases the fractal size by one peer and responds with its profile (identifier, overlay propagation service endpoint and leave date). When the new peer receives the profile, it adds it to the registry.

- **Registry Update (same ontology).** The bootstrapping procedure has the drawback that registries only contain entries for peers that are online at a given time. This is why every time a peer receives a gossiping message, it looks into the header, extracts the path and updates the registry with any peers that are not already in the registry. Every time a peer comes back online, peers that have been online in the meantime will have a better view of the system. The peer therefore checks the advertised registry of the fractal and adjusts the fractal size in its copy of the registry. Given the characteristics of the gossiping protocol, peers do not need to know the exact size of the fractal. The number of target peers and gossiping rounds (TTL) grows with logarithm of fractal size and do not change significantly with small changes in fractal size.
- **Registry Update (new ontology).** In some cases, ontologies affecting the overlay may change. Such changes affect the organization of the fractals and may lead to the addition of new fractals or the merging of existing ones. In the first case, the system needs to create a new registry for each new fractal. In the second case, some fractals in the current view merge with the fractals in the updated view. In both cases, the system needs to process the core ontology and create semantic connections (equivalence links) between the old concepts and the new ones (see next section). If a peer realizes that the topology has changed (reflecting a change in the core ontology), it automatically converts the old resource semantic types to the new ones, using the connections defined in the new ontology. From this moment on, it gossips to the new fractal.

Message Propagation in Semantic Overlay

When a peer receives a message, it first checks its resource semantic type and *validates* it with the corresponding ontology (see Registry Update above). The peer then calculates the group G—its partial view of the system. To do this, it analyzes the relationships between the different fractals belonging to same dimension.

Consider the following example.

The VDI represents a publication about Action and Romantic Movies (suppose we have one fractal for each one of these genres) and also addresses peers that understand concepts about Hunting and Kisses (two fractals belonging to another dimension). The final group of peers will thus satisfy the relationship:

$$G = (Action \cup Romantic) \cap (Hunting \cup Kissing)$$

Fig. 4.6 The target group of peers G

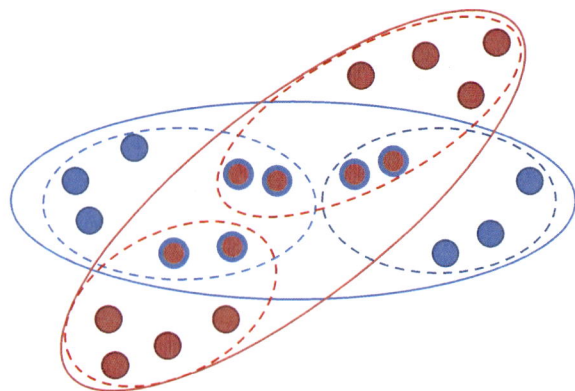

This is illustrated in Fig. 4.6, where the group of peers G consists of the peers marked with both blue and red colour.

Each peer then randomly selects $\log|G|$ other peers from its partial view, sends to them the Publication or the Subscription VDI, and stops. This implies that peers do not send data at every gossiping round—only once every time they receive a Publication or a Subscription VDI. This procedure is repeated by each peer for R rounds, where:

$$R = \frac{\log|G|}{\log \log|G|}$$

Each peer that receives a Publication or a Subscription VDI during a gossip round stores it in local buffers. These VDIs will be later used for possible matching. On each round, the peer reads its buffers, reduces R by one, selects $\log|G|$ peers uniformly at random and sends the VDIs to these peers using their publication or subscription service endpoints, given in the registry. The time between rounds depends on the trade-off we want between the delay of the VDI propagation and the resources' utilization.

In brief:

- When a peer receives a Publication or a Subscription VDI, it stores it in its buffers. The VDIs that have not been gossiped by this peer yet are marked as *ready to go*;
- When the time for gossiping has come, the peer inspects its local buffers for VDIs marked as *ready to go* and:
  - determines the partial view G;
  - selects peers to gossip to;
  - reduces the gossip message TTL (initially, TTL <=> Rounds);
  - sends the gossip message (including the VDI) to the selected peers.

4.3 Publish-Subscribe Pattern

Description

Publish and Subscribe operations are carried out at the middleware level of the
CONVERGENCE system. From the middleware perspective, publication of a
Resource VDI involves the following main steps:

1. Creation of a Publication VDI containing:

   a. Link to the stored Resource VDI (mandatory);
   b. VDI Metadata, usually taken from the Resource VDI (mandatory);
   c. Licence regulating access to the publication (optional);
   d. Event Report Request defining reports to be issued when specific events
      occur (optional);

2. Injection of the Publication VDI into the overlay of peers forming the "discovery overlay";
3. Storage of the Publication VDI on the network.

Symmetrically, the subscription process involves:

1. Creation of a "Subscription VDI" containing:

   a. One or more representations of the semantic subscription to a set of meta-
      data, in the form of a SPARQ query or a list of requested metadata
      (mandatory);
   b. Licence regulating access to the subscription (optional);
   c. Event Report Request defining reports to be issued when specific events
      occur (mandatory)";

2. Injection of the Subscription VDI into the overlay of peers forming the "discovery overlay";
3. Storage of the Subscription VDI on the network.

This procedure enables searching for subscriptions using standard Search
Content procedures, and easy matching of publications to subscriptions. Users are
therefore given the possibility of revoking their subscriptions and publications, and
of extending/altering them at a later time, as with any other VDI. Users need only
to create and submit to CONVERGENCE new corresponding VDIs that supersede
the previously injected VDIs. By using VDIs to carry subscriptions and publications, CONVERGENCE is able to exploit the Event Reporting mechanisms
defined in MPEG-21 part 15 (ISO/IEC 21000-15), embedding Event Report
Requests (ERRs) into each Subscription and Publication VDI. Subscription and
Publication VDIs are uniquely identified, and carry the requested metadata plus a
reference to the address of the "home" peer of the user to be notified if a match is
found. The reference can take the form of a peer identifier or a User VDI.

The use of specific licenses for Subscription VDIs and sequences of Sub-scription VDIs, enable many possible extensions. For instance licenses could restrict the kind of information that can be subscribed to by a particular class of users or they could support focused "forgetting of subscriptions", so that some subscriptions would immediately disappear from the system if not satisfied, while others would persist indefinitely. The mechanisms to define licenses in CONVERGENCE are described in Chap. 7.

The subscribe operation is carried out by invoking a Subscribe Content Service. Similarly, the publish operation is carried out by a Publish Content Service. Both are complex operations that involve a chain of Elementary Services. They are thus Aggregated Services. This implies there is no simple Publish Content ES, or Subscribe Content ES, but rather a pub/sub workflow involving coordination of middleware engines.

CONVERGENCE publication and subscription operations fully comply with the "content based" publish/subscribe paradigm as defined in (Eugster et al. 2003) and support structured semantic descriptions of content. This is much better than "topic based" approaches that require subscribers to know the full name of the content they require.

Semantic subscription makes it possible to store requests for events that change the state of a certain specific VDI (e.g. creation of a new version of the VDI, revocation of the VDI, the creation of a link to the VDI). They also allow sub-scribers to subscribe to VDIs that have not yet been published e.g. VDIs for:

- New models that will be produced by John;
- Special sales of mobile phones that will occur at a local store;
- Providing associations that will be made on columns of a certain author;
- Future releases of movies of a favourite actor;
- Collecting statistics concerning the type of manipulation certain content has been subjected to by other users.

In CONVERGENCE the two cases above are equivalent from the operational point of view and are thus treated using the same generic procedures.

When users create a subscription they issue a (semantic) query. The system then returns a list of VDIs that were already published in CONVERGENCE and that matched the query. The user can then select the most relevant VDIs from this matching list and/or wait for results that will be announced later. This requires a system wide request that is universally understandable. It also shows one of CONVERGENCE's "Unique Selling Points".

In existing systems, when a user subscribes to a particular kind of CD (e.g. Iron Maiden CDs) in a music retailing service (say music.com), he/she is informed of the CDs only when they are advertised in music.com. In CONVERGENCE the subscription is system wide and the user receives a notification every time anyone publishes information that is related with the CDs the user is interested in.

Conceptually, the subscription process is split into three parts:

- Part 1: inserting a semantic subscription system wide;
- Part 2: matching a subscription once relevant content is published;
- Part 3: delivering a notification to the subscriber.

Part 1: inserting/storing a semantic subscription system wide

Let us consider the following: We have two users: P (who publishes) and S (who subscribes). Let us also consider the case in which the subscription is prior to the publication of material matching the subscription.

When S subscribes, he/she provides semantic metadata, possibly organized as a complex query to filter out unwanted content and to allow grouping and sorting the results. These metadata are semantically validated by the CDS, which helps the user to formulate the query correctly (see relevant paragraphs about the CDS of this same deliverable, Sect. 4.5).

S may optionally provide a licence governing the use of the information contained in the Subscription VDI.

Along with these metadata, an Event Report Request is created. The ERR defines the event type that will trigger the notification (we call it a "Match" event type) and additional information needed to locate the peer that will receive the notification.

This ERR is inserted into the Subscription VDI. As mentioned earlier, it includes so called "DeliveryParams" (see ISO 21000-15), which is the address of the peer to be notified if the event occurs. According to the Event Reporting standard this can be an URI, a CoNet network resource (or service) name, a reference to a User VDI, an email address, etc.

Therefore a Subscription VDI will contain:

1. A structured description of the metadata of the resource the subscriber is interested in;
2. A Licence (defining who can do what with the subscription);
3. An Event Report Request containing:

   a. The verb "Match" applied to this Subscription Metadata and any corresponding Publish Metadata;
   b. The Identifier of the Device or User receiving the Event Report.

This set of information is published to the system by injecting its content into the overlay and storing the full Subscription VDI on the network. The same concepts and mechanisms are used for publication.

Part 2: matching a subscription

Later on, when P publishes content, he/she also provides the (semantic) metadata that best describes the published resource. The metadata are then injected into the overlay system.

When the gossiped information arrives at the destination peers, the procedure is reversed and all pending semantic search operations are performed on these peers, using those data as a target. If a subscription that matches the target metadata is found (such as the previous subscription of user S), it becomes a match candidate. So, whenever a Publication VDI reaches a peer, at each publish, it is evaluated against outstanding semantic queries from Subscription VDIs known to the peer.

With this approach, a rendezvous node between Subscription and Publication VDIs in the peers responsible for the common Resource Semantic Type is created.

Part 3: delivering matches to subscribers.
Each match between a new publication and an existing subscription generates an Event Report (ER), thus directly satisfying the request made in the matched Subscription VDI in the form of an embedded ERR. The ER is delivered to the peer/user referenced in the matched Subscription VDI, using an appropriate transport protocol. Different alternatives can be used, as for example invoking the CoNet *SendTo* primitive, or sending an E-mail message). When the user receives the notification, or more specifically an Application working on the user's behalf, the matching VDIs can be directly retrieved using the Deliver Content protocol.

This solution makes use of the MPEG-21 Event Reporting standard, in conjunction with the concept of a Subscription VDI. In this way, the MPEG-21 event notification mechanism is separated from the internals of injecting metadata to peers. If required, the regular MPEG-21 event notification mechanism can still operate at the client–server level, within the scope of a custom application. To apply the MPEG-21 event notification mechanism system-wide we use specific Subscription VDIs to inject events. This requires the creation of a new verb ("Match") in the Rights Data Dictionary (see ISO 21000-6). Details of this extension to the MPEG-21 standard are described in Chap. 7.

4.4 Semantic Search and Content Matching

This section provides a detailed explanation of the implementation of semantic search in CONVERGENCE. As in other P2P search overlay structures, the adopted mechanism is fundamentally asynchronous. However CONVERGENCE's publish/subscribe paradigm provides additional decoupling in time and space. Search requests are carried in VDIs. An expiry date dictates whether the user wishes for an immediate reply (as in present-day search engines) or is willing to accept results that may come later, in the next hours or days. Aggregation of similar results is carried out by the event report service that notifies the user of relevant matches. Below we provide a detailed walkthrough of the process.

A CONVERGENCE-compliant Application (i.e., an application that is able to invoke the functionality of the CONVERGENCE middleware) is running on a random PeerX, as for example a public device in the city hall, or the user's iPad. The application uses services and technologies offered by the CoMid.

- User UserA, types in some metadata (e.g. resource MOVIE, and genres SCI_FI and CRIME), related to resources he/she wants to subscribe to.

1. The application invokes the CDS.

   - The CDS expands the request by finding appropriate concepts in known domain ontologies. For example, if the user or the search application on his/her behalf, accepts the IMDB Mapping Movie Ontology (Grimnes 2012), the CDS maps the user request to the IMDB:Movie and IMDB:Genre classes;
   - The CDS formulates a correct semantic query that reflects the subscription. For example "SELECT ?x WHERE {?x a IMDB:Movie. ?x IMDB:genres "SCI-FI". ?x IMDB:genres "CRIME"}", where the prefix IMDB: qualifies the IMDB ontology;
   - The CDS helps the user find one (or more) Resource Semantic Types within the CONVERGENCE Core Ontology, and possibly within other semantic dimensions appropriate for the subscription (see chapter on Overlay TE, section Overlay TE.), For example, the user might select the cco:VIDEO semantic type, where cco: is the prefix of the CONVERGENCE Core ontology. In this way, the user exploits the concept of fractals and focuses the search on peers that are interested in the cco: VIDEO type of content (peers that have previously published or subscribed to cco: VIDEO content and have actively joined the fractal). The CDS prepares an incomplete metadata tag describing the Resource Semantic Type of the new Subscription VDI (the VDI identifier is still not known).

2. The application invokes Create Content.

   - A Subscription VDI is created which contains:
     - A unique identifier (SVDI_1);
     - The completed metadata indicating the core semantic type of the VDI, i.e. an RDF triple such as {SVDI_1 cco:hasReST cco:VIDEO}, embedded inside a didl:Descriptor tag;
     - The requested query, inserted into another didl: descriptor tag;
     - Licence and ERR;
     - Expiry date of the VDI.
   - The ERR says "notify peer PeerA if match". PeerA is the "home" peer of UserA. It is a peer that will receive the match notifications (the Event Reports).PeerA runs the Store Event and the Request Event servers "of the user".

3. The application invokes Inject Content (see Fig. 4.7).

   - If PeerX is not yet a member of the cco:VIDEO fractal, it registers to it and joins;
   - Information about the Subscription VDI is circulated in the fractal. Thus, the Subscription VDI SVDI_1 reaches the peers that participate in the overlay

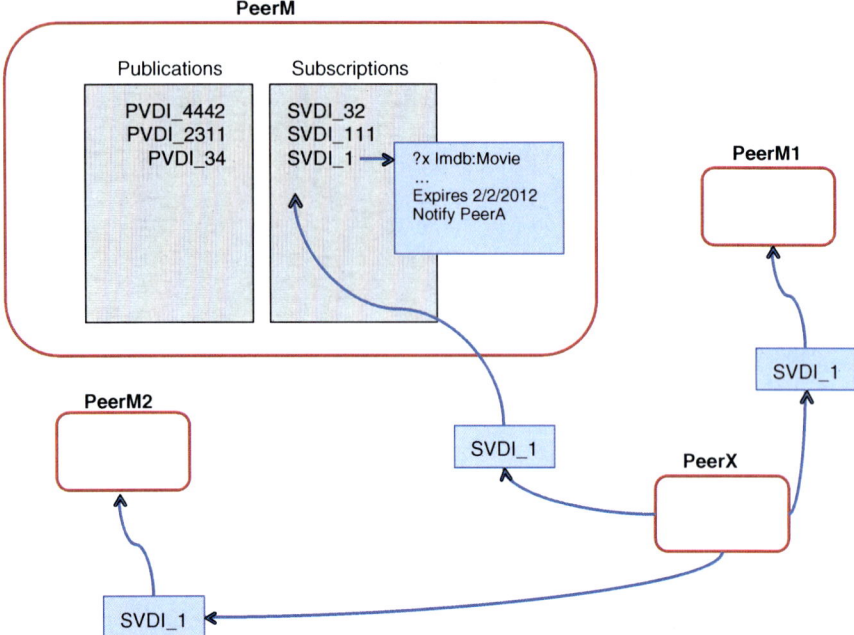

Fig. 4.7 Semantic subscription SVDI_1 reaches PeerM, PeerM1 and PeerM2 of the VIDEO fractal

fractal assigned to it, i.e. the fractal named "VIDEO" (a fully qualified name would be in the form urn:overlay:registry:cco:video);

- Each peer extracts information contained within the Subscription VDI (the identifier of the VDI, the embedded query and any additional metadata, license, ERR) and copies it to its own "Subscriptions Table". The "Sub-scriptions Table" keeps track of all Subscription VDIs that reach the peer, and indexes them using their VDI identifier.

4. The application invokes Store Content.

- The Subscription VDI is stored by CoNet as a generic network resource (just like any other VDI). This means that knowledge about the subscription is no longer restricted to the CoMid and can be **made available to crawlers that are not based on CoMid** (subject to security restrictions enforced by CoNet).

Another user makes a publication at a later time. That is: on a random PeerY (see Fig. 4.8):

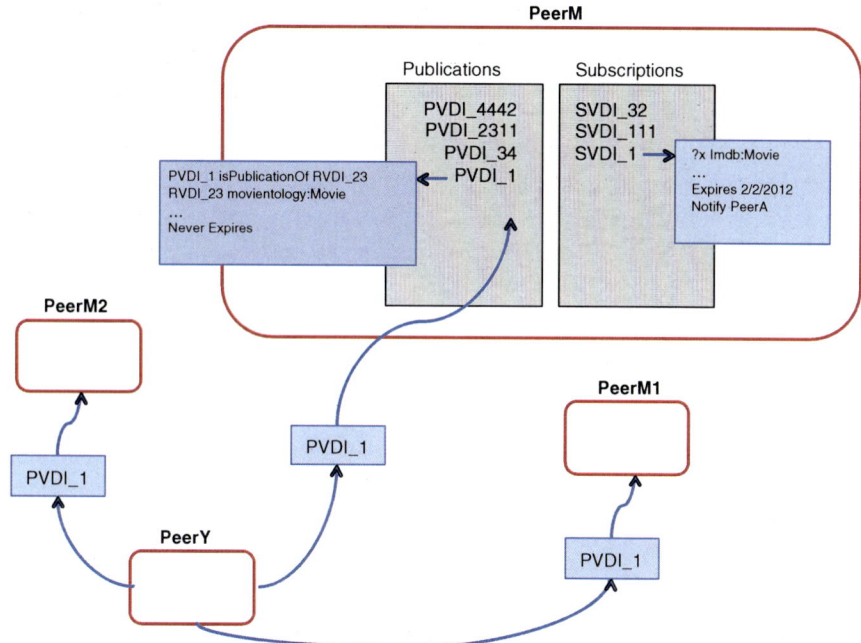

Fig. 4.8 Publication PVDI_1 reaches PeerM, PeerM1 and PeerM2 of the VIDEO fractal

5. A user decides to publish a resource (e.g. a movie) for which he (or somebody else) has already created a Resource VDI. The Resource VDI describes the movie and the license that regulates access to it. The Publication VDI copies those Descriptors and adds an expiry date and license. Metadata describing the resource is collected from its Resource VDI, and possibly refined by the user. The user now has the role of a publisher. We assume the author of the resource has given her permission to publish it.

- RDF triples describing the movie are extracted from its Resource VDI RVDI_23. The movie has been described using an alternative domain ontology about movies (e.g. the Movie Ontology at www.movieontology.org), which focuses on a specialized taxonomy of movie genres);
- The following RDF Descriptor tags, taken from the original Resource VDI, are inserted into the new Publication VDI:

  - {RVDI_23 rdf:type movientology: movie}
  - {RVDI_23 movientology:belongsToGenre movientology: Sci-Fi}
  - {RVDI_23 movientology:belongsToGenre movientology: Thrilling}
  - {RVDI_23 movientology:belongsToGenre movientology: actionreach}

The subject of the above semantic relationships is obviously the original resource VDI. They are now inserted into another VDI. This possibility is ensured by CONVERGENCE's concept of semantic links between VDIs. A specific tag allows RDF fragments to describe a VDI. In the case of publications, such VDIs simply state that they carry information about other VDIs.

6. The application invokes the CDS.

   • The CDS helps the user find one (or more) Resource Semantic Types within the CONVERGENCE Core Ontology or other shared semantic dimensions that provide appropriate classifications for the publication. For example the user selects the cco: VIDEO semantic type.

7. The application invokes Create Content.

   • A Publication VDI is created that contains:

     – A unique identifier (PVDI_1);
     – The original didl: descriptor metadata tags, now "describing" the Publication VDI;
     – An RDF metadata triple describing the core semantic type of the Publication VDI and thus indicating the Publication VDI's destination fractal in the overlay (this information is embedded in another didl: descriptor);
     – An explicit reference to the Resource VDI the Publication VDI refers to, i.e. {PVDI_1 cco:isPublicationOfRVDI_23};
     – Optionally a licence and an ERR, if the publisher wants to limit discoverability of the publication or to be notified when the publication is matched;
     – Expiry date of the VDI.

8. The application invokes Inject Content.

   • If PeerY is not yet part of the cco: VIDEO fractal, it registers to it and joins
   • Information about the Publication VDI is circulated: PVDI_1 reaches the peers that participate in the overlay fractal assigned to it, i.e. the "VIDEO" fractal.
   • Each peer extracts information contained in the Publication VDI (the identifier of the VDI, all metadata, license, ERR) and copies it to its own "Publications" table. The "Publications" table keeps track of all Publication VDIs that reach the peer, and indexes them using their VDI identifiers.

9. The application invokes Store Content.

   • The Subscription VDI is stored by CoNet as a generic network resource, just like any other VDI. This means that knowledge about the subscription is no longer restricted to the CoMid and can be made available to crawlers that are not based on CoMid (subject to security restrictions enforced by CoNet). The publication and the subscription VDIs reach several rendez-vous peers of the VIDEO fractal. This is illustrated in Fig. 4.9.

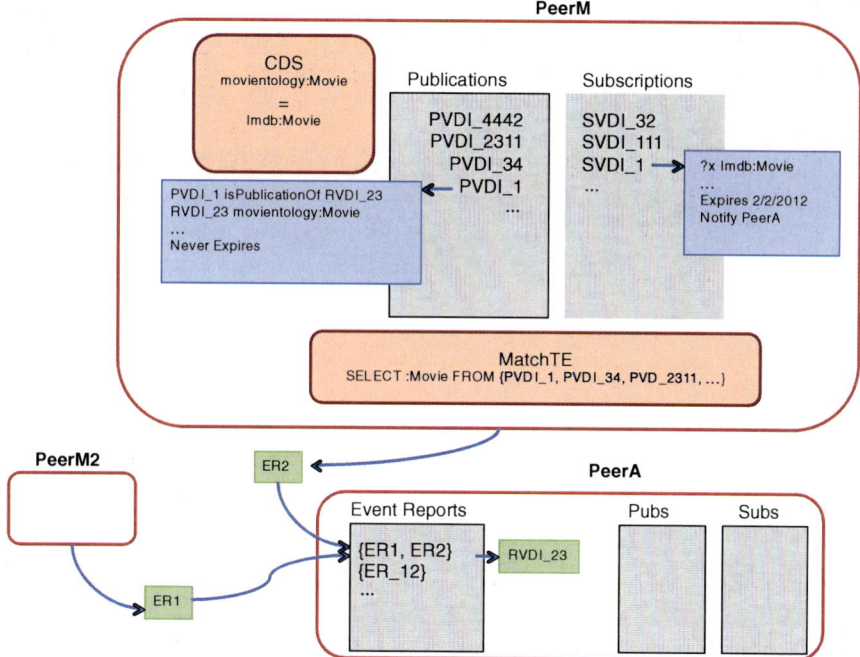

Fig. 4.9 Subscription is matched by PeerM and PeerM1

10. Match TE constantly monitors peer tables, being triggered on arrival of a
    Publication VDI or a Subscription VDI. The goal of the Match TE is to
    perform a match between the SPARQL queries embedded inside the Sub-
    scription VDIs stored in the "Subscriptions" table, and the metadata stored in
    the peer's "Publications table".

    • When a Publication VDI arrives at the peer, the full list of SPARQL queries in
      the "Subscriptions" table is scanned, and run against the newly arrived
      metadata;
    • When a Subscription VDI arrives at the peer, the embedded SPARQL query is
      run against the full list of metadata found in the "Publications table".

Before attempting to match a SPARQL query against the available data, the
system evaluates possible semantic equivalences between concepts belonging to
different ontologies. This is accomplished through the execution of the following
steps:

• Match TE invokes CDS, which asserts semantic equivalences between: the IMDB:
  movie **class** and the movieontology: movie **class**; the movientology: belong-
  sToGenre **property** and the IMDB:genres **property**; and the movientology: Sci-Fi

individual of type movientology:Genre and the "Sci-Fi" **literal** assigned to the IMDB:genres property.[3] The task of finding equivalent genres is greatly facilitated by the preliminary assertion of the equivalence of associated properties (preceding bullet), and of course by the matching of the literal with the individual name.

If more than one SPARQL query is to be executed the Match TE runs a query engine capable of processing semantic queries. The target data for the query is the aggregate of all metadata relevant to the query, after query expansion of the semantic classes. Match TE asks what are the RDF triples that satisfy the query and fetches them.

If matching RDF triples are found, a "Match Event" is triggered. The ERR is extracted from the Subscription VDI, the address of PeerA is read and companion ER is generated.

The ER is filled with the identifiers of all Resource VDIs known to PeerM, which contain matching metadata RDF triples.

11. Match TE invokes Event Report TE

   • The ER is sent to PeerA using the transport protocol specified in the original ERR, and invoking PeerA's Store Event service.

12. Since PeerA may receive several match notifications from different peers of the fractal, the peer performs a "notification fusion" eliminating duplicate VDI ids matching the subscription.

   When the user wants to check whether his/her home peer has received new notifications, he/she polls the Request Event service of PeerA from PeerX, where he/she is currently located. Alternatively, if PeerA is the user's mobile/laptop, a GUI alert pops up when the ER is received.

4.5 In Conclusion

CoMid adopts and extends the MPEG-M standard to provide the functionality required to the higher layers for the implementation of a semantic-based publish-subscribe model for the future Internet.

Besides the fact that its core operation relies on the use of open standards, thus promoting interoperability and widen its scope of application, CoMid can also be considered as a very flexible and efficient solution as it offers system-wide operation at the semantic level, being agnostic to the specific networking solution adopted. Additionally, by adopting and extending the MPEG-21 standard it is able

---

[3] The IMDB ontology contains a detailed taxonomy of genres in the form of a class hierarchy. The Movieontology allows representation of genres through string literals.

to manipulate in an integrated way the various types of information related to a single resource or entity, providing also the means to automatically control/govern the way that information is used.

This chapter described the semantic mechanisms developed by CONVERGENCE and how they are used together with standardized engines and technology, notably from MPEG-M and MPEG-21, to support the semantic publish-subscribe paradigm. Metadata is manipulated within a content-aware semantic overlay, running on top of the networking platform, namely on top of the CONVERGENCE content-centric network (CoNet).

Semantic interoperability is achieved through the use of a CDS, constituting an added value of the approach, as it promotes a system-wide operation where queries and interests of users are universally understood and served. The CDS, on one side, helps users to formulate their queries and express their interests by suggesting concepts contained in custom/user ontologies or in well-known domain ontologies; on the other side, it promotes the harmonization across different descriptive schemes during the search process, as it is able to establish equivalences between entities in different ontologies.

A flexible and efficient implementation of the publish-subscribe paradigm is obtained by using the VDI structure to inject user subscriptions into the overlay. The MPEG-21 Event Report Request mechanism is transparently embedded into the user subscription VDI, automatically triggering the generation of a corresponding Event Report when the interests of the user (as expressed in the subscription VDI) are matched to already published resources. This report is sent automatically to the subscribed user, thus notifying him/her of the availability of useful content. Additionally, given the possibility of defining and associating usage rights to each VDI, users can impose restrictions on the way their subscriptions or publications are manipulated by the system and consumed by other users.

References

A. J. Ganesh, A. M. Kermarrec, L. Massoulié, Scamp: Peer-to-peer lightweight membership service for large-scale group communication, in: J. Crowcroft, M. Hofmann (Eds.), Networked Group Communication, volume 2233 of Lecture Notes in Computer Science, Springer Berlin/Heidelberg, 2001, pp. 44–55.

A. M. Kermarrec, L. Massoulié, A. J. Ganesh, Probabilistic reliable dissemination in large-scale systems, IEEE Transactions on Parallel and Distributed Systems 14 (2003) 248–258.

B. Mandelbrot, The Fractal Geometry of Nature, W.H. Freeman and Company, New York, 1982.

G. Grimnes, The IMDB Mapping Movie Ontology, http://www.tecweb.inf.puc-rio.br/hyperde/export/309/branches/hyperde_danielle/db/IMDB-rdfs.rdf, Last accessed 26 November 2012.

ISO/IEC 21000-15—Information technology—Multimedia framework (MPEG-21)—Part 15: Event Reporting.

ISO/IEC 21000-6—Information technology—Multimedia framework (MPEG-21)—Part 6: Rights Data Dictionary.

M. Muhr, R. Kern, M. Granitzer, Analysis of structural relationships for hierarchical cluster labeling, in: Proceeding of the 33rd international ACM SIGIR conference on Research and development in information retrieval, SIGIR'10, ACM, New York, NY, USA, 2010, pp. 178–185.

ISO/IEC 23006-4—Information technology—Multimedia Service Platform Technologies—Part 4—Elementary Services.

P.T. Eugster, P.A. Felber, R. Guerraoui, A. Kermarrec, The Many Faces of Publish-Subscribe, in ACM Computing Surveys (CSUR), Volume 35, Issue 2, 2003.

P. T. Eugster, R. Guerraoui, A.-M. Kermarrec, L. Massouli'e, Epidemic information dissemination in distributed systems, IEEE Computer 37 (2004) 60–67.

R. Friedman, D. Gavidia, L. Rodrigues, A. C. Viana, S. Voulgaris, Gossiping on manets: the beauty and the beast, SIGOPS Operating Systems Review 41 (2007) 67–74.

Chapter 5
The Versatile Digital Item

Helder Castro, Angelo Difino, Giuseppe Tropea
and Nicola Blefari Melazzi

Abstract This chapter provides the definition of the Versatile Digital Item, the basic unit for data distribution used within the CONVERGENCE system. It explains how the VDI builds on, and extends the scope, of the MPEG-21 Digital Item to build a self-contained data package that can be used to encapsulate any kind of digital information in an information-centric, publish-subscribe framework. This chapter details some of the most relevant aspects pertaining to the structure of the VDI, its identification, its connection into sequences, and its logical interweaving into a fabric of inter-VDI relationships. It also explores some implications of the above aspects on the system's operations.

5.1 Introduction

The VDI is the basic unit for data distribution and transaction within the CONVERGENCE system. Upon it fall a number of requirements that pertain to the framework's operational nature which focuses on a content centered and publish subscribe paradigm.

The VDI is defined in accordance with the MPEG-21 standard. It builds on it, to attain self-containment, authenticity insurance, unique identification and inter-VDI relationship expressiveness.

H. Castro (✉)
INESC TEC, Faculty of Engineering, University of Porto, Porto, Portugal
e-mail: hcastro@inescporto.pt

A. Difino
CEDEO.net, Turin, Italy

G. Tropea · N. Blefari Melazzi
Electronic Engineering Department, University of Rome "Tor Vergata", Rome, Italy

F. Almeida et al. (eds.), *Enhancing the Internet with the CONVERGENCE System*,
Signals and Communication Technology, DOI: 10.1007/978-1-4471-5373-3_5,
© Springer-Verlag London 2014

We begin with a focus on the VDIs context. Thus, in Sect. 5.3, we present the most relevant aspects of the VDI's base standard (MPEG-21 2003, 2005), focusing on the concept of Digital Item. In Sect. 5.3 we explain what a VDI precisely is and in Sect. 5.4 we detail its lifecycle by looking into such phases as the VDI's creation, publication, relationships establishment and removal. Section 5.5 presents the main conditioning factors that guide the conception of the VDI's specifics.

We then go on to focus on the VDIs internal details. We present its structure in Sect. 5.6. Some of the next sections focus on specific relevant aspects of the VDIs structure and contents. Section 5.7 explains the employed scheme for the unique universal identification of VDIs across the CONVERGENCE framework, and for their inscription into VDI sequences. Section 5.8 extensively approaches a number of key aspects related to the declaration of inter-VDI and inter-Item relationships. Section 5.9 reflects about the placement of the actual resources within the VDI structure. Section 5.10 focuses on the granularity of the VDI, whereas Sect. 5.11 approaches trust related issues pertaining to the VDIs.

Finally Sect. 5.12 focuses on the implications that specific operations, on VDIs, have on the VDIs structure and on the system's operational stability as a whole.

5.2 What is a Digital Item?

The MPEG-21 Digital Item is the basic unit of data transaction in the MPEG-21 framework. A Digital Item is a combination of resources (video, audio tracks, images, etc.), metadata (creator identification, resource category, etc..), event monitoring pertaining instructions and intellectual property information (license, encryption resource information).

In MPEG-21 the representation of a Digital Item is performed through the employment of a set of XML elements, structured in accordance to the MPEG-21 standard (MPEG-21 2005). An example of such a structure is depicted in Fig. 5.1 and its most relevant composing elements are described in the sub-sections bellow.

5.2.1 Item

An *Item* is a grouping of sub-*Item* and/or *Component* elements, These are bound to relevant descriptive information, which is carried by *Descriptor* elements. Items may contain options, which allow them to be customized or configured. Items may also be conditional (on predicates asserted by selections defined in explicitly declared choices). An *Item* that contains no sub-*Items* can be considered an entity—a logically indivisible work. An *Item* that does contain sub-*Items* can be considered a compilation—a work composed of potentially independent sub-parts. *Items* may also contain annotations to their sub-parts.

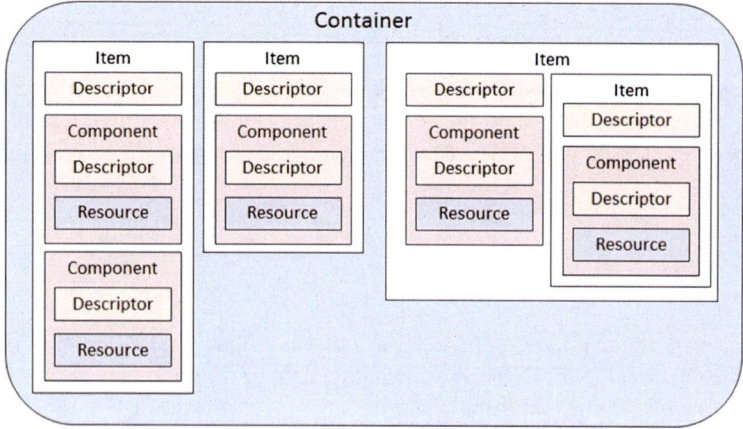

Fig. 5.1 DID structure example

The relationship between *Items* and Digital Items (as defined in (MPEG-21 2005) could be stated as follows: *Items* are declarative representations of Digital Items.

5.2.2 Component

A *Component* is the binding of a resource to relevant descriptive information, pertaining to it, by way of *Descriptor* elements. Such *Descriptor* elements bind the mentioned information to all or part of the specific resource instance. They will typically contain control or structural information about the resource (such as bit rate, character set, start points or encryption information), but not information describing the semantics of the "content" within.

It should be noted that a component itself is not a Digital Item; components are building blocks for items.

5.2.3 Resource

A *Resource* element represents an individually identifiable object such as a video or audio clip, an image, or a portion of text content. A *Resource* may also, potentially, represent a physical object. All resources represented by *Resource* elements are locatable via an unambiguous address.

5.2.4 Descriptor

A *Descriptor* element associates information with the enclosing element. This information may be specified in the form of a *Component* element (constituting, for instance, a thumbnail of an image, or a text component), or a textual statement.

5.2.5 Statement

A *Statement* element carries a literal textual value that contains information, but not an asset. Examples of likely statements include descriptive, control, revision tracking or identifying information.

5.2.6 Identification

The Digital Item Identification (MPEG-21 2003) standard defines the syntax and semantics of the Digital Item *Identifier* and *Related Identifier* elements. These elements make it possible to uniquely identify a Digital Item and to associate identifiers that are related to the Digital Item, container, component, and/or fragment thereof, which do not identify the Digital Item directly. These identifiers will be in the form of a URI.

5.2.6.1 Identifier

Digital Items and their parts, within the MPEG-21 Multimedia Framework, are identified by encapsulating Uniform Resource Identifiers (URI) in the *Identifier* element.
Syntax:

```
<xsd:element name="Identifier" type="xsd:anyURI"/>
```

Semantics:
This element contains an identifier for a Digital Item, container, component, and/or fragment thereof in the form of a URI. A Registration Authority maintains a list of identification schemes compliant with ISO/IEC 21000-3. Identifiers are not required to be registered with the Registration Authority to be conformant to this clause of the specification.

This definition of the Digital Item *Identifier* element is the basis for the definition of the VDI identifier adopted in CONVERGENCE. Each VDI has a unique identifier, the VDI ID, which follows the above definition of the DII standard.

While the MPEG standard dictates that if you change a single bit of a VDI you must change the VDI identifier, the *Related Identifier* element offers great flexibility in associating multiple identifiers to the same Digital Item when they are needed for different purposes.

5.2.6.2 Related Identifier

While the *Identifier* element enables the unique identification of Digital Items (or parts thereof), the *Related Identifier* element allows the identification of information that is related to the Digital Item (or parts thereof).

Syntax:

```
<xsd:element name="Related Identifier" type="xsd:anyURI"/>
```

Semantics:

The *Related Identifier* element may not be used for identifying the Digital Item (or part thereof). This shall be done by using the Digital Item *Identifier* element as specified above.

A Registration Authority maintains a list of identification schemes to be used within ISO/IEC 21000-3. Identifiers are not required to be registered with the Registration Authority to be conformant to this clause of this specification.

The Digital Item Identification MPEG standard contains an example, copied in the following XML snippet, which clarifies the use of this element. The example shows how to uniquely identify a resource (an MPEG Audio Layer III encoded sound track), within a Digital Item using an International Standard Recording Code (ISRC). The example also highlights how to associate a related identifier (here: identifying the underlying music work with an International Standard Work Code (ISWC)) with such a resource.

Having multiple *Resource* elements embedded inside the same *Component* element means that the various represented resources are totally equivalent, and an agent may choose any one of them. On the other hand, having multiple *Components* grouped inside one *Item* means that the represented Digital Item is composed of several different pieces of information. Moreover, each *Component* can have its own *Descriptors* that provide additional information about the represented digital component.

```
<?xml version="1.0"?>
<DIDL
xmlns="urn:mpeg:mpeg21:2002:01-DIDL-NS"
xmlns:dii="urn:mpeg:mpeg21:2002:01-DII-NS">
<Item>
<Component>
<Descriptor>
<Statement mimeType="text/xml">
<dii:Identifier>
urn:mpegRA:mpeg21:dii:isrc:US-ZO3-99-32476
</dii:Identifier>
<!– ISRC identifying the sound recording –>
</Statement>
</Descriptor>
<Descriptor>
<Statement mimeType="text/xml">
<dii:RelatedIdentifier>
urn:mpegRA:mpeg21:dii:iswc:T-034.524.680-1
</dii:RelatedIdentifier>
<!- ISWC identifying the underlying musical work ->
</Statement>
</Descriptor>
<Resource ref="Track01.mp3" mimeType="audio/mp3"/>
</Component>
</Item>
</DIDL>
```

The structure of the DI also allows for an *Item* element to be composed of a sequence of sub-*Items*. So what is the difference between having multiple *Items* and having multiple *Components* inside an embedding *Item*? The answer is provided by the very definition of the MPEG-21 standard, where it states that *Items* are intended to be the lowest level of granularity visible to an end-user. This means that a user interface could allow end-users to access the *Item* children within an *Item* element, but not the *Component* children within an *Item* element.

5.2.7 Example of a Digital Item

In the XML snippet presented below, an example is given of a Digital Item consisting of a compilation of musical tracks.

```
<?xml version="1.0"?>
<DIDL xmlns="urn:mpeg:mpeg21:2002:02-DIDL-NS"
xmlns:dii="urn:mpeg:mpeg21:2002:01-DII-NS">
<Container>
<Item id="Audio Compilation">
<Descriptor>
<Statement mimeType="text/xml">
<dii:Identifier > myaudiocompilation:1 </dii:Identifier>
</Statement>
</Descriptor>
<Descriptor>
<Statement mimeType= "text/plain">
The compilation of my favourite songs
</Statement>
</Descriptor>
<Item id ="Track 1">
<Descriptor>
<Statement mimeType ="text/plain">
Track one of the compilation
</Statement>
</Descriptor>
<Component>
<Resource ref=TrackOne.mp3" mimeType ="audio/mpeg"/>
</Component>
</Item>
<Item id="Track 2">
<Descriptor>
<Statement mimeType ="text/plain">
Track two of the compilation
</Statement>
</Descriptor>
<Component>
<Resource ref="TrackTwo.mp3" mimeType ="audio/mpeg"/>
</Component>
</Item>
</Item>
</Container>
</DIDL>
```

The overall Digital Item is represented by the top level Item element, which is named as "audio compilation". It represents the entire set of resources that comprises the compilation. The top level Item contains one or more inner Items, termed "content element items". Each such Item represents one specific part of the overall Digital Item, i.e. an audio track.

5.3 What is a VDI

A VDI is an extension of the MPEG-21 Digital Item concept and of its profile that was designed within the context of MPEG-M. It is thus a general purpose container that may be employed to declare any virtual or physical object. Such objects may be media resources, services, people and Real World Objects (RWOs).

A VDI, like an MPEG-21 DI, binds metadata (describing the structure, content and context of an item) and its corresponding resources (such as other VDIs, audio, images, video, text, descriptors of RWOs and people etc.).

The metadata comprises information pertaining several such aspects as: the secure, location independent and universal identification of the VDI; the cryptographic protection and authentication of the VDI's content; the management of the rights of the parties involved in the manipulation/transaction of the VDI; the VDI's expiry date so as to support its "digital forgetting"; the semantic description of the VDI's content; the description of the relationships between the VDI and other VDIs; etc.

The VDI thus functions as a location independent multipurpose container of information which is stored and distributed throughout the CONVERGENCE framework, based on a publish-subscribe paradigm. End-users and applications are then able to, depending on their individual rights, create, publish, search for, retrieve, consume, tag, subscribe and "unpublish" VDIs.

5.4 VDI Life Cycle

In Fig. 5.2 we depict the main stages of a VDI's lifecycle. The typical life cycle of a VDI, thus, consists of the following fundamental phases:

- Creation—the original production of the VDI by its creator user;
- Publication—the divulging, for public knowledge, of the VDI's existence and of the semantics of its contents;
- Interconnection—the establishment of logical connections, from other VDIs, to it;
- Removal—the removal of the VDI from the public scope.

Furthermore, given the publish-subscribe paradigm that shapes CONVERGENCE's operation, all the events that take place within the publication, interconnection, or removal phases, generate notifications to subscriber users. In this

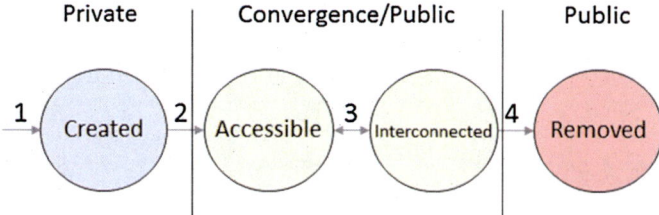

Fig. 5.2 The VDI's life cycle

way subscribers can find out when a targeted VDI has been published, modified (updated with new version, interconnected) or unpublished.

The VDI is considered to be private until it is published on the CONVER-GENCE framework. From that moment on, the VDI will be in the public sphere, where the CONVERGENCE system, and its users, may act on it, within the limits defined by its access rights. Once it is unpublished, the VDI will once again become visible only to the publisher.

The main actors involved in the life of a VDI are:

- The Creator—creates and manages the VDI throughout its life span;
- The User—searches for, subscribes to, visualizes, interconnects, and creates new versions of the VDI according to her access rights;
- CONVERGENCE system—stores, retrieves and manages the VDI while it is accessible to the public.

The next sub sections present the above mentioned phases in greater detail.

5.4.1 VDI Creation

In this phase, a user packages a specific group of resources into a new VDI. To do so, the creator:

1. defines the metadata for the VDI;
2. attaches resource(s);
3. includes and/or links any additional attachments and/or other VDIs;
4. chooses an identification scheme and requests a Content Identification Service to assign the VDI's identifier;
5. defines the license and signs the VDI.

The VDI is then created in the user's personal space, containing structure described in Table 5.1.

Table 5.1 Structure of a VDI

VDI content	Source	Details
Creator	Creator	Details about the user who created the VDI
Searchable metadata	Creator	Ontology instances, keywords
Resources	Creator	Links to resources or embedded resources
Relationships	Creator	Semantic references to other VDIs
Security data	Creator/system	Creation date, expiry date, license, digital signature
VDI identifier	Creator/system	Unique identifier for the VDI
Status data	System	Created
History	System	Links to previous versions of VDI

Table 5.2 Propagation of a VDI

VDI content	Source	Details
Creator	Creator	Details about the user who created the VDI
Searchable metadata	Creator	Ontology instances, keywords
Resources	Creator	Links to or embedded resources
Relationships	Creator	Semantic references to other VDIs
Security data	Creator/system	Creation date, expiry date, license, digital signature
VDI identifier	Creator/system	Unique identifier of VDI
Status data	System	Created
History	System	Links to previous versions of VDI

5.4.2 VDI Publishing

In this phase the VDI moves from the private sphere onto the CONVERGENCE realm. The creator publishes the VDI, onto CONVERGENCE, by invoking the relevant CoMid functionalities to gossip and advertise the VDI's content in CoMid and to store it in CoNet, thus making it accessible to the public.

The VDI is thus ready to:

- be discovered, in searches, retrieved and consumed by authorized users (according to the creator defined access rights);
- be found by user subscriptions;
- generate notifications when specific events occur.

From this point on, some of the information contained in the VDI is propagated through the CONVERGENCE framework. Table 5.2 highlights said information in boldface.

Table 5.3 Permanent VDI fields

VDI content	Source	Details
Creator	**Creator**	**Details about the user who created the VDI**
Searchable metadata	Creator	Ontologies instances, keywords
Resources	Creator	Links to or embedded resources
Relationships	Creator	Semantic references to other VDIs
Security data	Creator/system	Creation date, expiry date, license, digital signature
VDI identifier	Creator/system	Unique identifier of VDI
Status data	System	Created
History	**System**	**Links to previous versions of VDI**

5.4.3 VDI Interconnection

This is the most active phase in a VDI's life cycle. Throughout its lifespan a VDI may be interwoven into a complex fabric of inter-VDI semantic relationships. This fabric is built by the creator community as it produces and publishes further VDIs which maintain some logical relationship with it (e.g. VDIs containing corrections, responses, extensions, etc., to it), and adequately declare it within their descriptive metadata.

Said declaration may take the form of a versioning declaration or of an inter-VDI relationships declaration. These mechanisms are described in detail in later sections.

Overall, a VDI can be changed through a number of actions:

- the publisher modifies the VDI;
- users, with the appropriate access rights, modify the VDI;
- users associate other VDIs to it (through links, comments, extensions, etc.).

Each change produces a new version of the VDI, which, in its descriptive metadata, links back to its previous versions. Newer versions of a VDI, replacing older ones, may hold completely different information. However some fields (e.g., Creator, History), remain the same. In Table 5.3, these fields are highlighted in boldface.

5.4.4 VDI Removal

This phase marks the end of a VDI's lifecycle The VDI is removed from the CONVERGENCE framework, and is no longer available to the public. This event occurs when:

- the VDI expires;
- the publisher revokes the VDI.

Table 5.4 Information affected by the removal of a VDI

VDI content	Source	Details
Creator	Creator	Details about the user who created the VDI
Searchable metadata	Creator	Ontologies instances, keywords
Resources	Creator	Links to or embedded resources
Relationships	Creator	Semantic references to other VDIs
Security data	Creator/system	Creation date, expiry date, license, digital signature
VDI identifier	Creator/system	Unique identifier of VDI
Status data	System	**Expired/revoked**
History	System	Links to previous versions of VDI

Following a VDI's removal, the status data, pertaining to it, is updated, across the framework, to specify that the VDI is inactive, and the reason for said inactivity (expiry or revocation). The affected information is highlighted in boldface in Table 5.4.

5.5 VDI Design Guidelines

Input to the VDI design process comes from three main sources: user-driven requirements, technical-driven requirements and coherence with existing standards. The above requirements were condensed into three main design guidelines, in what pertains to VDIs: attaining self-containment; allowing the explicit formation of logical sequences and enabling the explicit establishment of inter-VDI logical relationships.

5.5.1 Self-containedness

In the CONVERGENCE network, IP addresses are replaced by resource names, and such resources are retrieved or accessed on a content-centric manner, as opposed to the present location-centric manner. This approach enables the optimization of the network's use and of content accessibility and has implications on the design of VDIs. CONVERGENCE thus places the emphasis on the "self-contained" nature of the VDI, as it is this feature that allows the physical nodes, participating in a CONVERGENCE-enabled communication network, to make routing decisions based on the VDI itself (and on the network identifier), without the need for any external information.

5.5.2 Sequence Forming

The design of the CONVERGENCE system takes into account requirements that were identified after an extensive analysis of COVERGENCE's context of operation and of the panorama of expectations and needs that it presents. In this regard, the members of the CONVERGENCE consortium have devised a number of real-world use cases. Alongside, their requirements and technical implications were coherently coalesced into CONVERGENCE requirements.

All devised scenarios require a functionality that is perceived by end-users as an update of the VDI. From a user's perspective, this is very natural: all web sites are inherently dynamic in nature—e.g. the home page of a big site will change from hour to hour even though the name of the site remains constant. The mentioned usage scenarios identified many situations where the need for such functionality arises. The following are some such situations:

- Changing the ownership field in a VDI representing an appliance;
- Changing the "user manual" resource in a VDI representing an appliance;
- Adding information about a safety recall in a VDI representing an appliance;
- Modifying a comment to a photo, where the comment is carried in a "comment VDI" and the photo in "photo VDI";
- Changing access rights to a video represented by a "video VDI".

As a VDI may ultimately be updated various times, and the earlier versions may still be desired, sequences of logically preceding VDIs will be formed. Attaining the possibility to update VDIs, and thus build VDI sequences, is, therefore, a key design guideline for the CONVERGENCE system.

5.5.3 Relationship Expression

Several literature studies including (Tsinaraki et al. 2007; Bekaert et al. 2003; Rodriguez-Doncel and Delgado 2009); have proposed that Digital Item relationships should be described within DIs by means of their explicit declaration within DIDL documents. Typically, that information should be introduced by the DI creators. Such a declaration is a powerful, yet lightweight, way of defining a semantically rich relational fabric pervading all VDIs, which will greatly ease the automatic "interpretation" of VDIs and their context.

In terms of CONVERGENCE use cases, this strategy makes it easier to specify that some VDIs are:

- comments to other VDIs—e.g. a *VDI C* "carrying" a comment to *VDI V* which "carries" a movie;
- responses to other VDIs—e.g. a *VDI R* "carrying" a response to a *VDI Q* which "carries" some specific question;

- enhancements of other VDIs—a *VDI 2* which "carries" the latest version of a software whose earlier version is "carried" by *VDI 1*;
- additions to other VDIs—a *VDI X* "containing" an X-Ray image of Mary's lungs is an addition to a *VDI MR* which "contains" Mary's overall medical record;
- etc.

The introduction of the explicit expression of relationships between VDIs enables consumer users to explicitly search for comments, corrections, evaluations, related to some specific other VDI. This enables CONVERGENCE to offer a richer content searching service. This strategy ensures that the identification of inter-related VDIs is a strait forward process and does not depend on the system somehow identifying relationships though the semantic tags used to qualify the content (a complex and inevitably imprecise task).

For instance, if *VDI A* "contains" a movie and *VDI B* "contains" a review and there is no explicit declaration of the inter-VDIs relationship, the only way for CONVERGENCE to determine that the latter is a review of the former would be by examining the semantic tags of both VDIs. These would have to explicitly indicate that the resource represented by *VDI B* is logically a review of the content represented by *VDI A*. Otherwise the only way to automatically establish the relationship would be through complicated and imprecise inferential mechanisms.

In brief, the lack of formal, well-defined mechanisms to specify relationships amongst VDIs, would make it difficult to provide CONVERGENCE users with effective search services. However, if *VDI B* carried metadata, explicitly declaring that it maintains a relationship of "reviewing" with *VDI A*, then, CONVERGENCE can offer a service allowing users to explicitly search for reviews of a specific VDI.

Attaining the capability to unambiguously capture the above mentioned relations is thus a design guideline for CONVERGENCE and its VDIs.

5.5.4 Addressing the Guidelines

The flexibility provided by the MPEG-21 standard for Digital Item Declaration (MPEG-21 2005) and Digital Item Identification (MPEG-21 2003), makes it possible to extend the DI concept into a self-contained VDI object, capable of supporting the requirements of a content-centric and convergent Internet, the formation of VDI sequences and the expression of VDI relationships. Said flexibility is enabled by the possibility to perform the "overloading" of the MPEG-21 *Descriptor*. MPEG-21 It is thus an optimal tool for the definition of VDI objects.

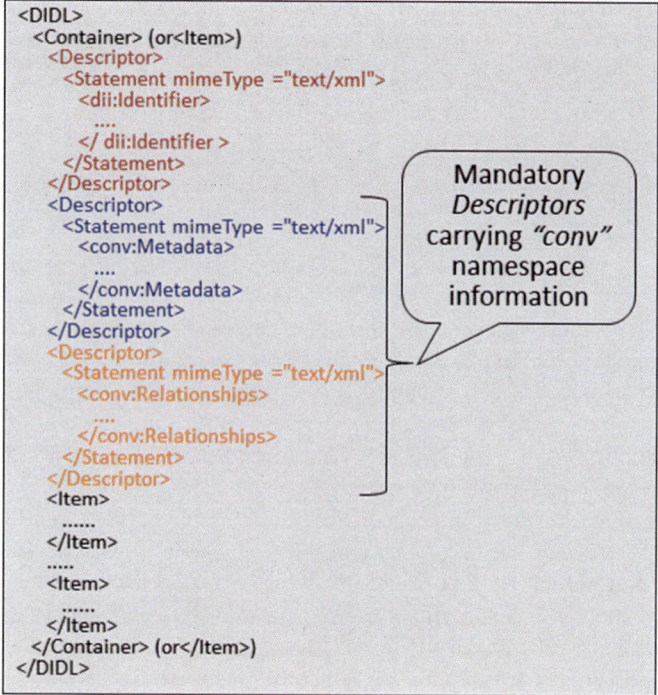

```
<DIDL>
  <Container> (or<Item>)
    <Descriptor>
      <Statement mimeType ="text/xml">
        <dii:Identifier>
        ....
        </ dii:Identifier >
      </Statement>
    </Descriptor>
    <Descriptor>
      <Statement mimeType ="text/xml">
        <conv:Metadata>
        ....
        </conv:Metadata>
      </Statement>
    </Descriptor>
    <Descriptor>
      <Statement mimeType ="text/xml">
        <conv:Relationships>
        ....
        </conv:Relationships>
      </Statement>
    </Descriptor>
    <Item>
    .......
    </Item>
    .....
    <Item>
    .......
    </Item>
  </Container> (or</Item>)
</DIDL>
```

Mandatory *Descriptors* carrying *"conv"* namespace information

Fig. 5.3 Conceptual representation of a VDI structure

5.6 Structure of the VDI

The CONVERGENCE VDI is a fully compliant and fully conformant MPEG-21 Digital Item DIDL document, with additional required fields to convey:

- A unique VDI identifier—a mandatory unique identifier at the root level of the VDI document (generically referred to as VDI_ID from now on), which is assigned, to the VDI, when it is first created;
- A declaration of the VDI's relationships with other VDIs—describing the VDIs relational context, including its versioning of other VDIs;
- VDI semantic characterization metadata—searchable information containing a precise semantic characterization of the VDI's content.

Formally, a VDI may thus be defined as a DI whose root element (either a *Container* or an *Item*), embeds mandatory *Descriptors*, as presented in Fig. 5.3.

Conceptually, the VDI is thus an enrichment of the root element of a DI with a set of *Descriptor* elements, which are used to wrap the mandatory VDI identifier, the declaration of relationships between the VDI and other VDIs (capturing, among other things, the versioning information), and the semantic metadata describing the VDI's content.

The VDI identification is expressed employing the (MPEG-21 2003) *Identifier* element, while the CONVERGENCE self-defined "*conv*" namespace identifies and isolates the *Relationships* and *Metadata* blocks.

The VDI creation process could be conceptually seen as a:

- definition of a "regular" DI;
- embedding identification, and "*conv*" specific, information in the root element.

An existing DI can thus be turned into a VDI without opening or un-wrapping it, if this is not necessary for other reasons. All that is necessary is to add *Descriptor* fields to the root element in order to identify it uniquely in the CONVERGENCE system and to express the necessary semantic relationships. No other manipulation of the existing elements is required.

Unfolding of the nested structure is necessary only if semantic relationships, pertaining to the nested elements, are to be expressed. In such a case the nested Items, composing the VDI, will also be enriched with relationship descriptions. Otherwise they can be left untouched.

As explained, in previous chapters, MPEG-21 Digital Items can contain sub-Items, and thus, *Item* elements may carry inner *Item* elements. This implies some kind of "nesting" or "composition" relationship between the sub-Items and the embedding Item. Since we are including, in the VDI, the means to explicitly express semantic relationships between Items, via the overloading of *Descriptors*, it is thus possible to define semantic relationships for nested Items below the root level. This can be done by defining explicit semantic relationships between Items of the same VDI. This technique can be very useful for representing RWOs, as it enables the definition of the full range of relationships that may exist between the parts that compose the object, without relying on the rigid relationship expressing capabilities of nesting Items within Items.

5.7 VDI Identification

5.7.1 VDI Individual Identification

As previously mentioned, the MPEG-21 Digital Item Identification standard prescribes the *dii:Identifier* tag as part of *Descriptor* statements. This implies that there is no unique identifier for the whole DI: wherever a *Descriptor* can fit, there is also room for an identifier. In fact, each *Container*, each *Item* composing the DI, and even each *Component* can contain one (or more) *dii:Identifier* elements, embedded within their respective *Descriptor* elements. This implies that CONVERGENCE needs to define the syntax of its unique VDI identifiers and their expected position in the DI structure.

In terms of their syntax, individual VDI identifiers are built, as prescribed by MPEG-21, through the employment of *dii:Identifier* elements as the statements of

Descriptor elements (MPEG-21 2003, 2005). In what regards the placement of the identification information in a VDI, its mandatory place, as already implied in Sect. 5.6, is at the "top" of the VDI as a direct child of the root *Container* or *Item* element.

5.7.2 VDI Sequence Identification

5.7.2.1 Rationale

As stated earlier, VDI updating is a key requirement for the CONVERGENCE system. This challenges the design of the VDI as a self-contained package of data. MPEG standards dictate that each DI should be packaged and signed for purposes of trust, and that, if a single bit changes, a new DI with a different identifier must be issued. This means that CONVERGENCE should issue new VDIs for each update, never changing a signed VDI. However, it should also give users the perception that they are updating their VDIs.

This cannot be implemented by designing a VDI with a changeable part and an unchangeable part. This approach would invalidate the VDI's signature. In theory it might be possible for the original author to sign the unchangeable part and other actors to sign the parts that they introduced/changed. However such a scheme would present complex security problems that, even if solvable, would require considerable work and would lead to contrived architectural solutions.

What is, therefore, needed is a level of indirection. In the devised scheme, updating a VDI does not involve any modification of VDIs that have already been published (including modification to their individual identifiers), though it may lead to removal of obsolete VDIs. Said scheme translates into different features at the different levels of the CONVERGENCE architecture:

- VDIs will contain a sequence identifier (SEQ_ID), and additional metadata indicating their "ancestor" VDIs. This will make it possible to insert new VDIs, and to declare "ancestry" inter-VDI relationships (see Sect. 5.8) in internal fields. This functionality will be implemented at the CoMid level;
- All VDIs in a sequence will be advertised as CoNet level named-network-resources corresponding to their VDI_IDs. At the same time an updated named-network-resource will point to the latest version corresponding to the SEQ_ID. No VDIs will ever be altered. At most they will be deleted. Old VDI_IDs will not be reused in new VDIs. This chain of advertisement operations is to be implemented at the CoNet level.

From a logical point of view, these features will, naturally, require the extra VDI sequence identifier (which may be thought of as a sort of a VDI group identifier), and some additional metadata fields inside the VDI to express "ancestry" relationships. This will allow search and retrieval of the latest version of an information package or of any of the preceding versions. The stable sequence identification, besides the individual VDI_IDs, allows referring to the sequence as a whole.

In this way the VDI sequence can be seen as a container for evolving, dynamic information (a sequence evolving along some semantic axis, such as time, space, functionality, etc.).

5.7.2.2 Scheme

Whenever a VDI is created, which is not part of any existing sequence of VDIs, it is given a new, system wide unique, identifier (e.g. *VDIx-v1*) and a new sequence identifier (e.g. *VDIx*). The syntax for the sequence identifier is compatible with that of the VDI identifier, and with the MPEG-21 DII standard.

When a new VDI, belonging to the same sequence is to be added, (which typically constitutes a VDI update), that new VDI is also given a new unique identifier (e.g. *VDIx-v2*), but its sequence identifier is the same as that of VDI *VDIx-v1*, (i.e. *VDIx*).

From that point on it is no longer possible to refer to the updated VDI with the "old" VDI identifier (*VDIx-v1*). To access the "most up to date" version of a VDI sequence the VDI sequence identifier must be employed. To retrieve any specific VDI from a VDI sequence it must be referenced through it individual identifier (e.g. *VDIx-v1* or *VDIx-v2*).

As subsequent VDIs are added to a VDI sequence (presumably in update procedures), a chain of related VDIs is formed, as depicted in Fig. 5.4. Summarizing:

- each VDI instance has a VDI identifier and a VDI sequence identifier;
- end-users, CoMid, and VDIs themselves, can refer to VDIs using the VDI identifier (if the target VDI is a specific fixed point in the VDI sequence, which will, thus, never be updated), or a VDI sequence identifier (if the target VDI is the "most up to date" VDI in a VDI sequence);

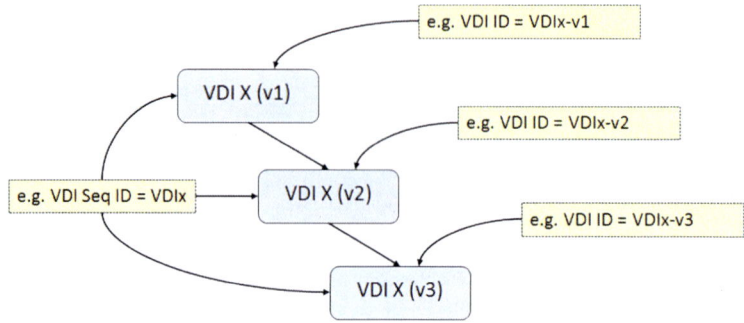

Fig. 5.4 Formation of VDI sequences

- "old" instances of a VDI will still be individually accessible (as each one has its own unique VDI identifier), until they are eliminated. If it is not necessary to maintain the older VDIs in a Sequence, the CoMid will delete references to the old copies. Trying to retrieve such older VDIs will then return a "not found" error.

5.7.3 Identification Scheme Across CONVERGENCE Layers

The CoMid needs to identify VDIs and VDI sequences. VDI identifiers and VDI sequence identifiers, are thus CoMid level identifiers. Inside the VDI, they are captured as *dii:Identifier* elements.

The CoNet needs to identify named resources. Its identifiers thus differentiate between such resources. However, the CoNet's IDs are also addresses. They must thus carry the necessary information to enable the retrieval of the information objects that they represent.

There is, thus, a need for identifiers at two different levels, and for the means to relate them to each other. To do so CONVERGENCE employs a global static mapping between CoMid and CoNet IDs, which dispenses the existence of any added translation service.

To attain said mapping, while enforcing a route-by-name strategy at the CoNet, the VDI identifier (CoMid level), must be composed of two parts:

1. a routing prefix—this is the part of the identifier that CoNet employs to locate the network principal which is responsible for the identified VDI;
2. an application specific identifier or label—this is the part of the identifier that differentiates it amongst all the VDIs made available by the previously mentioned principal.

Such VDI identifiers may then be used at the CoMid and CoNet levels.

5.8 VDI Relationships

5.8.1 What is a VDI Relationship

VDI relationships are logical links that connect two individual VDIs (or, more precisely, connect the resources represented by two individual VDIs), from the point of view of human consumers. As such, they describe the "meaning" of a referencing VDI with respect to a referenced VDI. In this relationship, the referencing VDI may be considered the active object in the relationship and the referenced VDI the passive object. Together with the relationship itself, the two VDIs form a semantic triple. For example, if *VDI A* "carries" a user-made video and the creator of *VDI A* expresses (in *VDI A*) his appreciation for another movie "contained" in *VDI B*, then *VDI A* (plays the active role and) is a comment to *VDI B* (which plays the passive role). That is, *VDI A* maintains a relationship of "*commentation*" with *VDI B*. If a user publishes a book "within" a *VDI C*, and later on she realizes that it contains errors, she may then publish a *VDI D* containing the book's errata. In this situation, *VDI D* maintains a relationship of correction with *VDI C*.

5.8.2 Relationship Declaration

As already explained, in CONVERGENCE VDIs, semantic relationships are defined via a custom overloading of the *Descriptor* element. This is accomplished by introducing a self-defined CONVERGENCE "*conv*" XML namespace, containing XML statements about VDIs. After considerable analysis of various possible syntactic solutions for the actual expression of relationships, we concluded that RDF/XML is the most adequate choice. Therefore, *conv:Relationships* elements shall carry snippets, expressed in the RDF (W3C 2004) language, based on a vocabulary defined in OWL (W3C 2012).

VDI relationships are thus declared as RDF metadata carried within specifically defined tags (*conv:Relationships* tags), contained inside VDI *Descriptor* elements. A VDI may declare an arbitrary number of relationships to an arbitrary number of other VDIs (as exemplified in Fig. 5.5).

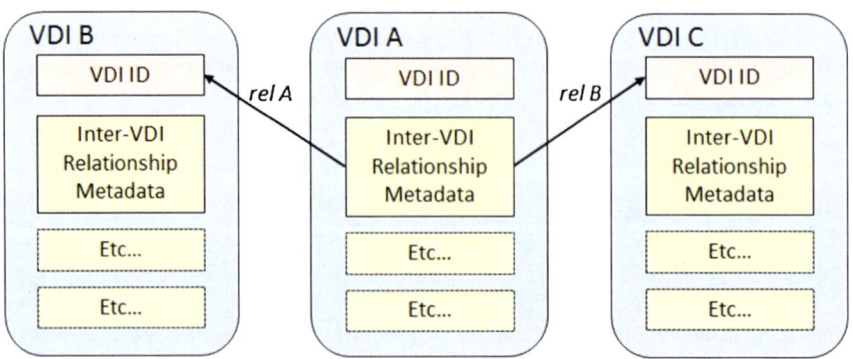

Fig. 5.5 Graphical depiction of inter-VDI relationships

```
<?xml version="1.0"?>
<!DOCTYPE rdf:RDF [
<!ENTITY owl "http://www.w3.org/2002/07/owl#" >
<!ENTITY xsd "http://www.w3.org/2001/XMLSchema#" >
<!ENTITY owl2xml "http://www.w3.org/2006/12/owl2-xml#" >
<!ENTITY rdfs "http://www.w3.org/2000/01/rdf-schema#">
<!ENTITY rdf "http://www.w3.org/1999/02/22-rdf-syntax-ns#" >
<!ENTITY convcoreRelOnt "conv:corerelationalontology" >
]>
<rdf:RDF xmlns="conv:corerelationalontology"
xml:base="conv:corerelationalontology"
xmlns:owl2xml="http://www.w3.org/2006/12/owl2-xml#"
xmlns: convcoreRelOnt=conv:corerelationalontology"
xmlns:xsd="http://www.w3.org/2001/XMLSchema#"
xmlns:rdfs="http://www.w3.org/2000/01/rdf-schema#"
xmlns:rdf="http://www.w3.org/1999/02/22-rdf-syntax-ns#"
xmlns:owl="http://www.w3.org/2002/07/owl#">
<owl:Ontology rdf:about="conv:corerelationalontology"/>
<!- Classes ->
<owl:Class rdf:about="conv:corerelationalontology;DigitalItem"/>
<owl:Class rdf:about="&owl;Thing"/>
<!-Object Properties ->
<owl:ObjectProperty
rdf:about="&convcoreRelOnt;interDIRelationship">
<rdfs:domain rdf:resource="&convcoreRelOnt;DigitalItem"/>
<rdfs:range rdf:resource="&convcoreRelOnt;DigitalItem"/>
</owl:ObjectProperty>
</rdf:RDF>
```

The types of declarable relationships are derived from a base set of global relationships that is defined in a core relationship ontology. This ontology can be extended by application-specific purposes, with application-specific relationships. Said core ontology is presented in the previous text box, and depicted in Fig. 5.6.

Fig. 5.6 Graphical depiction of the core ontology

5.8.3 Relationship Level or Type

Within CONVERGENCE VDIs, relationships are defined at two levels:

1. At the VDI level—expressing the set of relationships in which the VDI, as a whole, is involved;
2. At the Item level—expressing the set of relationships in which specific parts of the VDI are involved.

Both levels have a model that captures the distinction between Context Information and Representation Information, as defined in the OAIS Open Archival Information System (OAIS 2012) and in similar approaches:

- Context Information—describes how the object relates to its environment and other objects;
- Representation Information—describes the internal structure and hierarchy of the object.

VDI level relationships can be seen as inter-VDI relationships as they describe the relationships of the VDI, as a whole, with other VDIs. Such a relationship may, for instance, describe an "illustration" relationship between an image "carrying" VDI and a text "carrying" VDI.

Item level relationships may, generally, be described as intra-VDI relationships, as they typically describe the relationships between the inner parts of the VID. For example, the complex details of the physical structure of an RWO can be modeled by introducing semantic relationships between the Items composing the VDI that represents the whole object.

However, the later type of relationships may also describe the relationships of an inner part of a specific VDI with exterior entities.

Relationships between VDIs (VDI level relationships), are expressed in terms of the VDI-level ontology. This defines the attributes of a VDI as a whole, as well as core system concepts that are vital for the functioning of the CONVERGENCE framework. Nevertheless it is a dynamic ontology. Wherever useful, it is possible to add application-specific relationships to describe the domain specific behavior of VDIs used by a CONVERGENCE-enabled application. User searches and queries are expressed in terms of restrictions formulated over the terminology of the VDI-level ontology and its extensions.

Inter-Item (Item level), relationships are expressed in terms of application ontologies. These may take the form of a root Item-level ontology. These introduce completely custom taxonomies to capture the semantics of the internal structure of specific VDIs for precise application domains. However these ontologies are conceptually associated with a single VDI and/or domain, and users can only fully exploit them when they acquire the VDI and browse the individual Items contained in it.

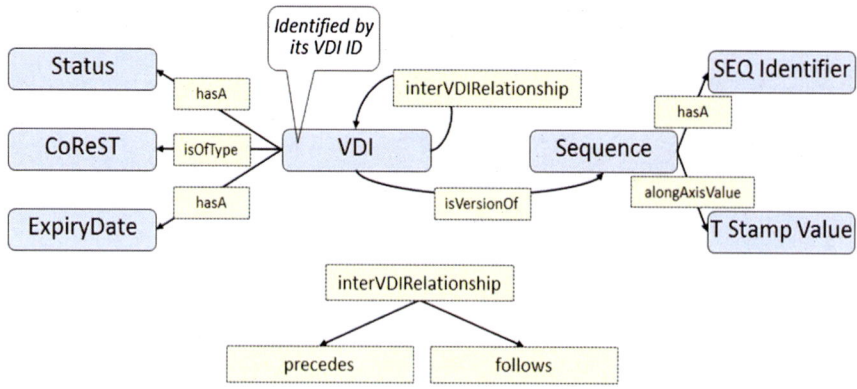

Fig. 5.7 VDI-level ontology model

5.8.3.1 VDI-Level Relationships

The basic taxonomy of possible relationships between VDIs and their meaning is encoded in the VDI-level ontology. Said ontology (which is built on the one presented in Fig. 5.6), is depicted in Fig. 5.7.

The purple-colored classes (centered on the VDI class) capture the Context Information for a VDI as a whole. Each VDI is assigned a CONVERGENCE Resource Semantic Type (*CoReST*). The explicit *Expiry Date* class is used to enforce "not found" behavior when searching for expired objects at the Content Level of the CONVERGENCE architecture, and for progressive garbage collecting at the Network Level.

Each VDI has a unique identifier, and is assigned to a *Sequence*, which represents a ternary relationship between a specific VDI, a global identifier for the whole sequence and a "semantic axis" along which the *Sequence* has to be interpreted. The default approach is to unfold the versioning along the time axis, so that each update to the sequence gets a progressive timestamp (class *T. Stamp Value*).

The yellow-colored areas (more specifically, the hierarchy under *interVDIRelationship* property), capture the Representation Information for a VDI as a whole, specializing the *inter VDI Relationship* generic bonding between two distinct VDIs. This captures the basic relationships for the declaration of VDI sequence structures, i.e. the "*follows*" and "*precedes*" relationships, which enable the specification of which VDIs follow which other VDIs in a VDI sequence.

Obviously, once the hierarchy under the *inter VDI Relationship* property is extended to support the declaration of other types of relationships (e.g. "*comments*", "*corrects*", "*extends*", etc.), it may then be employed, at VDI creation time, by creator users to, weave the VDIs into a rich relational fabric, which permeates VDIs as a whole, rather than the multiple individual resources within them. As implied before, that hierarchy may be freely extended by means of domain and application-specific terminology.

The semantically precise expression of VDI-level relationships makes it possible to combine searches, for instance, for VDIs which "comment" or "correct" a specific other VDI, with other search criteria such as such as sequence version, date of creation, geographic place of creation of the VDI, author identification etc.

5.8.3.2 Item-Level Relationships

The basic taxonomy of possible inter-Item relationships is encoded in the Item-level ontology. That ontology is depicted in Fig. 5.8. In said figure, the purple-colored class *Item* and the yellow-colored hierarchy of properties, (under *inter-Item Relationship*), distinguish, again, between Context and Representation information.

Since Item-level relationships apply to single Items within the VDI, each *Item* element must be uniquely identified, so that it may be unambiguously referred by the relationship declarations in question. The scheme employed to perform that identification is to use the global VDI identifier as a base reference (xml:base) and then employ standard XML fragment identification, namely the XPointer specification (XPOIN 2001; W3C 2001), for the differentiation between individual *Items*. This method has the advantage of being independent from other possible identifiers already present within the *Item*. Since existing *Item* elements are identified by their XPointer semantics within the VDI document, no explicit additional identifiers are necessary.

In every VDI the actual inter-Item relationships are expressed in terms of application specific ontologies. These extend the Item-level ontology by defining a terminology customized for the VDI's specific application domain. The taxonomy of possible relationships between parts of a VDI, represented by *Item* instances contained within a VDI document, is thus encoded in application ontologies.

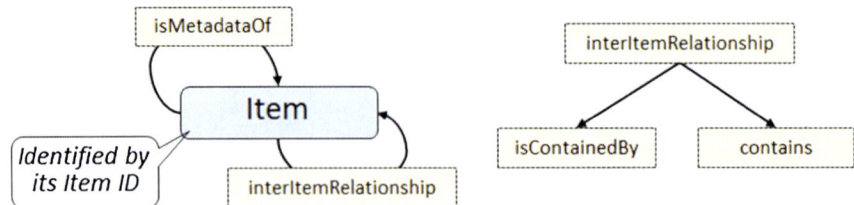

Fig. 5.8 Item-level ontology model

5.8.4 Implications of VDI Removal

In most Inter-VDI relationships, the removal of one of the VDIs in the relationship does not present problems. For instance, if VDI B comments VDI A and the latter is removed, the former will become a dangling comment that "points" to a non-existent VDI. In most cases this will not be a problem (just like, in most cases, a broken link does not disable a web site).

There are some cases, however, where a broken link could be problematic. This can occur when Inter-VDI relationships form logical sequences, where the knowledge of the entire chain is necessary for the proper operation of the system. The typical example of this are the updating Inter-VDI relationship chains (VDI C *updates* VDI B which *updates* VDI A), which are constituted as VDI sequences are built. In such cases, removal of one of the VDIs isolates the two resulting halves. This is why VDI sequences are managed by means of a common Sequence Identifier which is part of the context information of all VDIs.

The explicit declaration of Inter-VDI relationships, and the fact that sequence identification is independent from such declarations, thus helps the implementation of digital forgetting, making it possible to delete VDIs without impeding the regular operation of the system. It also facilitates digital forensics, as even if a specific VDI is removed, all the other VDIs that, for instance, commented and criticized it will still be available. This means that "traces" of the VDI will remain.

5.8.5 Relationships and Rights

A creator of a VDIs is not free to declare any kind of relationships, of its VDI, to other VDIs. For instance, if Mary is not the owner of *VDI A*, she is not allowed to insert a *VDI B*, which declares a *"correction"* relationship towards *VDI A*. On the other hand, Mary is perfectly free to insert a *VDI C*, which declares a *"commenting"* relationship towards *VDI A*. Before accepting a VDI the system has to analyze and validate its VDI-level relationship tags.

If a user searches for and retrieves *VDI C*, she only receives Mary's comment, not *VDI A* itself. Nevertheless when browsing *VDI C* she will be able to reach into *VDI A*, if this is allowed by the rights expressed by the later VDI.

5.8.6 Inter-VDI Relationship Declaration Example

In the text box below, an example of an inter-VDI relationship declaration is presented. More specifically it declares that a VDI identified as *"conv:vdi:A"*, maintains a relationship of *"enrichment"* towards another VDI identified as *"conv:vdi:B"*. That is, VDI *"conv:vdi:A"* logically enriches VDI *"conv:vdi:B"*.

```
.......
<Descriptor>
<Statement mimeType="text/xml">
<dii:Relationships>
<rdf:RDF xmlns="individual:example:ontology#"
xml:base="individual:example:ontology"
xmlns:owl2xml="http://www.w3.org/2006/12/owl2-xml#"
xmlns:extOnt = "sample:extension:ontology#"
xmlns:vdiLOnt="conv: vdiLOnt#"
xmlns:xsd="http://www.w3.org/2001/XMLSchema#"
xmlns:rdfs="http://www.w3.org/2000/01/rdf-schema#"
xmlns:rdf="http://www.w3.org/1999/02/22-rdf-syntax-ns#"
xmlns:owl = "http://www.w3.org/2002/07/owl#">
<owl:Ontology rdf:about="individual:example:ontology">
<owl:imports rdf:resource="sample:extension:ontology"/>
</owl:Ontology>
<owl:ObjectProperty rdf:about="&extOnt;enrichment"/>
<owl:Class rdf:about="& vdiLOnt;VDI"/>
<owl:Thing rdf:about="#conv:vdi:A">
<rdf:type rdf:resource="&vdiLOnt;VDI"/>
<extOnt:enrichment rdf:resource ="#conv:vdi:B"/>
</owl:Thing>
<owl:Thing rdf:about="#conv:vdi:B">
<rdf:type rdf:resource="& vdiLOnt;VDI"/>
</owl:Thing>
</rdf:RDF>
</dii:Relationships>
</Statement>
</Descriptor>
....
```

The above presented declaration presupposes the associated existence of an application specific ontology, referred to as ("*sample:extension:ontology*"), which, extending the core ontology depicted in Fig. 5.6, performs the definition of the "*enrichment*" relationship.

5.9 Resource Placement Within VDIs

The MPEG-21 standard enables resources to be placed within DIs in varied dispositions.

At a basic level, such resources may be referenced from the DI, or directly included in it. At a higher level, (the level of *Component* elements), for instance, multiple *Component* elements in the same *Item* element can be used to offer the same resource in different formats. For example two *Component* elements in the same *Item* element could hold files *song.mp3* and *song.wav* (i.e. the same song in compressed and raw formats). Furthermore, the same *Component* element can contain or reference multiple resources employing multiple *Resource* elements.

A DI may, for instance, combine the above possibilities to attain resource redundancy, by offering the same resource twice in the same *Component* element, once by value and once by reference (linking to copy of the resource stored at a remote server).

All that is stated above is also true for CONVERGENCE VDIs. However, although syntactically both the *Component* and the *Resource* elements can be tagged with *Descriptor* elements, within VDIs, the latter elements must not be used for embedding semantic relationships at the level of the *Component* or *Resource*.

5.10 Granularity of the VDI

Granularity is the degree to which information is packaged within the same VDI and the degree to which it is distributed across several distinct VDIs. Granularity is thus related to the minimum sized transacteable unit of information. The optimal granularity depends on the application domain. As defined in CONVERGENCE, the VDI is the atomic searchable entity of the CONVERGENCE network. All Items within the same VDI are classified under the same set of metadata within the CoMid and are distributed together as a self-contained information package.

5.11 Trusting VDIs

Every VDI is typically signed by its creator user. Furthermore, they may also contain a license. Such a license specifies the rights panorama pertaining to the VDI. It may state, for instance, that: only the authenticated author can modify (update) the VDI; other authors can add resources to the VDI (by updating it) without deleting the original resources; or that other authors can freely replace resources in the VDI (again through updates). In the first case, the author will sign any updated version. In the second case, the new author will add (and perhaps sign) a new resource, in the third case the author will replace the old resource with a new, signed resource.

In each of the above cases, the CoMid instance responsible for publishing the updated VDI, will recalculate a digital signature on the whole package of bits, so that the VDI can be trusted, based on the author of the change.

130   H. Castro et al.

Non-repudiation can be assured by a combined employment of the signing of
the VDI, by its author, and by further stating, in the license, that the VDI cannot be
modified by anyone (including the author). These authentication and governing
mechanisms are a powerful guarantee that the VDI has not been tampered with
after publication. CONVERGENCE also supports anonymous publication of
VDIs. This is essential, if we want to use VDIs for free discussion on controversial
topics or in adverse contexts.

5.12 Operating on VDIs

5.12.1 Processing Relationships at Packaging Time

The CoMid needs to be efficient. It is not efficient for CoMid instances to have to
run a SPARQL query, through the whole VDI, to collect its Metadata, every time
they receive a VDI. Metadata must thus be pre-packaged within VDIs. Therefore,
after its original creation and before it can be actually published to the CON-
VERGENCE framework, a VDI needs to be packaged for dissemination and
distribution.

Packaging operations mainly pertain to the processing of relationship declarations into *conv:Metadata* blocks. At VDI packaging time, the, CoMid level,
Packaging Engine scans the VDI, collects and interprets *Descriptor* elements for all
Items and normalizes the information contained in *conv:Metadata* elements, which
consists of RDF/XML data. Said data is thus packaged into a "normalized" form
allowing efficient parsing and searching. The normalization operations include:

- Creating an expanded list of triples;
- Resolving references;
- Making literal values explicit.

The metadata contained in the VDI's top *conv:Metadata* block pertains to the
VDI as a whole, and not just to some of its specific Items. It is obtained:

- from the *conv:Relationships* elements of all *Item* elements (i.e. Item-level
  relationship information);
- from the *conv:Relationships* element originally placed at root of the unpackaged
  VDI (i.e. VDI-level relationship information).

At publishing time, the CoMid parses the VDI, extracts the *conv:Metadata*
information, (which now contains the semantic characterization of the VDI and the
characterization of the relationships in which the VDI is involved), and performs
the appropriate dissemination and storage of that information. Users will thus be
able to search the data, both for specific concepts (keyword based search) and for
VDIs which maintain specific relationships with other VDIs (relationship based
search).

5.12.2 Processing Relationships at Search Time

In order to enforce user searches for VDIs the system will need to process the *conv:Metadata* blocks, which are disseminated at VDI publishing time. When a user searches for VDIs by keyword, the local CoMid instance sends the query to the peer CoMid instance which is responsible for the concept in question (that is, the CoMid peer that keeps the list of all the VDIs involved as active or passive part in that concept), which then sends the appropriate response back.

When, a user searches for VDIs which are involved in a specific relationship with another VDI (for instance, all VDIs that *"comment"* on VDI XYZ), the CoMid collective will search through the information describing relationships of the type in question, looking for *"commenting"* relationships which link to VDI XYZ. To perform such a search, the user hosting CoMid instance directs the query to the CoMid instance which is responsible for the relationship concept *"comment"*, that is, the CoMid peer that keeps the list of all VDIs involved in a relationship of that specific type. The peer in question then provides the appropriate answer. In this manner, the system handles the two types of search in the same way, that is, it directs the query to the appropriate CoMid instance.

5.12.3 VDI Updating Operations

Updating a VDI, as explained in Sect. 5.7.2, means that a new VDI will be added to the same VDI sequence (same sequence as the one that the updated VDI is in). The new VDI will thus become the "head" of the sequence, logically replacing the previous one.

The "head" VDI may have much of the same metadata as the previous one and thus, it may reutilize the latter's *conv:Metadata* blocks. The new VDI, though, will typically have to be newly signed by its creator user.

The publication of an updating VDI implies CoMid and CoNet level operations which are described in the next sub-sections.

5.12.3.1 VDI Sequences and CoMid Level Operations

A VDI update operation is mapped onto a CoMid level VDI insertion (for instance, of a new VDI with a brand new ID, but the same SEQ_ID as an earlier VDI, and a different timestamp). If necessary, the old VDI is removed. When the new VDI's metadata are gossiped, they are linked to the new VDI_ID. In this way, when the system finds a match for a user search query, the corresponding VDI is given back to the user through its link-back ID.

Each new VDI, in a VDI sequence, can, in theory, be completely different from its preceding VDI. During inter-CoMid "gossiping" interactions, the different CoMid instances, running at different peers, include the inter-VDI sequence identifier, SEQ_ID, in the information that they "gossip about". As a result, information propagates through the network and can be retrieved during the processing of user requests.

In this situation, an application can request from the CoMid:

- A specific VDI, defined by its VDI_ID;
- The latest version (or update) of a specific VDI sequence, identified by its SEQ_ID.

In the later situation the CoMid uses the SEQ_ID to identify the "top" VDI and to retrieve it.

5.12.3.2 VDI Sequences and CoNet Level Operations

To implement the functionality described in the previous section, each VDI update operation triggers the following operations at the CoNet level:

- the updating VDI is advertised, in the CoNet, with a NID that is the same as the SEQ_ID, in this way effectively replacing the old content;
- the updating VDI is also advertised in CoNet with the NID that is the same as the VDI_ID. This makes it possible to fetch VDIs either through their VDI_IDs or their SEQ_IDs.

Once an update operation is completed, the concerned SEQ_ID is thus present in more than one VDI. The network will thus contain, for instance, the original VDI, identified as http://vdi.org/VDIX and a new VDI identified as http://vdi.org/ VDIY. Both VDIs will share the same SEQ_ID, with a value, for instance, of http://vdi.org/seqXseqX. When a request is made for http://vdi.org/seqXseqX, the system fetches VDI Y, the most recent advertised VDI with that identifier.

In this way the CoNet level is "aware" of all of the update operations and keeps a reference to the most up to date version of a VDI.

5.13 Conclusions

The defined VDI object is indeed a versatile object. While strictly adhering to the MPEG-21 standard, it extends the MPEG-21 DI in order to render it a suitable vehicle for data transaction in a content oriented network.

The VDI is thus a self-contained object that is uniquely and universally identified across the CoMid and CoNet layers. It has the capability to be logically grouped into VDI sequences and is enriched with the capacity to declare its relational context. The VDI also comprises the necessary provisions to enable the expression of its usage rights, the preservation of its integrity and the validation of its authenticity across its life cycle.

Furthermore, the structure devised for the VDI, together with the specific characteristics of the CONVERGENCE system, enables also the maintenance of VDI authorship anonymity. Additionally, it becomes possible the enforcing of digital forgetting, without hampering the system's operation.

All of the above mentioned characteristics of the VDI are in line with its purpose of facilitating and empowering CONVERGENCE's operation in accordance with the publish-subscribe paradigm.

References

Chrisa Tsinaraki, Panagiotis Polydoros, and Stavros Christodoulakis, Interoperability Support between MPEG-7/21 and OWL in DS-MIRF, IEEE Transactions on Knowledge and Data Engineering, 19(2): 219–232, 2007.

Jeroen Bekaert, Patrick Hochstenbach, and Herbert Van de Sompel, Using MPEG-21 DIDL to Represent Complex Digital Objects in the Los Alamos National Laboratory Digital Library, D-Lib Magazine, 9(11): 2003.

MPEG-21: ISO/IEC, Information technology—Multimedia framework (MPEG-21)—Part 2: Digital Item Declaration, 2005.

MPEG-21: ISO/IEC Information technology—Multimedia framework (MPEG-21)—Part 3: Digital Item Identification, 2003.

OAIS, Reference model for an open archival information system, 2012. Available from: http://public.ccsds.org/publications/archive/650x0m2.pdf (accessed 18 January 2013).

Victor Rodriguez-Doncel, and Jaime Delgado, A Media Value Chain Ontology for MPEG-21, IEEE Multimedia, 16(4): 44–51, 2009.

XPOIN, XML Pointer Language (XPointer) Version 1.0, 2001. Available from: http://www.w3.org/TR/WD-xptr (accessed 27 January 2013).

W3C, Resource Description Framework, 2004, http://www.w3.org/RDF.

W3C, Web Ontology Language 2, 2012, http://www.w3.org/TR/owl2-overview/.

W3C, XML Pointer Language, 2001, http://www.w3.org/TR/WD-xptr.

Chapter 6
The CONVERGENCE Security Infrastructure

Thomas Huebner, Andreas Kohlos, Amit Shrestha and Carsten Rust

Abstract This chapter describes the Convergence security infrastructure. The core component for Convergence Security (CoSec) has a distributed architecture. It encompasses subcomponents on different computing platforms such as client computers and smart cards, application servers and peers. An essential feature of Co-Sec is the use of smart cards as a secure token. This Convergence token provides sensitive security functions on a tamper-resistant device. The chapter first introduces the concepts and the architecture of the security infrastructure. Based on a description of the basic cryptographic primitives, as well as of the advanced cryptographic schemes applied by the project, we describe the high-level security functions provided for the Convergence middleware and network layer.

6.1 Introduction

In the architectural design of Convergence, CoSec is the component in the Computing Platform level responsible for handling cryptographic protocols and other security-related tasks. CoSec has a distributed architecture encompassing several independent and possibly distant components with each component comprising software as well as hardware. In the application flow of security protocols, these components interact with each other. As a consequence, their APIs can be quite complex.

CoSec offers security support mainly to CoMid, the middleware layer of Convergence. It provides for instance encryption of content, authentication of users, signing of VDIs etc. It offers also similar support to CoNet, e.g. high-level functions for utilizing a hybrid encryption scheme in distributing media content

T. Huebner · A. Kohlos · A. Shrestha · C. Rust (✉)
Morpho Cards, Paderborn, Germany
e-mail: carsten.rust@morpho.com

F. Almeida et al. (eds.), *Enhancing the Internet with the CONVERGENCE System*,
Signals and Communication Technology, DOI: 10.1007/978-1-4471-5373-3_6,
© Springer-Verlag London 2014

over an Information-centric network. In CoMid, the security services implemented by CoSec are provided to all CoMid engines through the Security TE.

The Security TE on the one hand exposes high level functions as described in the previous paragraph, thereby allowing for abstraction from the CoSec cryptographic functionality (such as key-generation, random-number generation). On the other hand, these low-level functions are also included in the MPEG API for the Security TE, so that they are accessible for higher level engines and applications.

The cryptographic operations of CoSec must be executed in a trustworthy environment. In theory, we could assume that the entire network is safe, and that registered devices (laptops, PCs, etc.), once checked, will remain secure devices and can be trusted. In reality, however, networks can be anything but secure, and devices (even if initially correct) can be tampered with by fraudulent users, or manipulated by third parties.

Many of these problems can be solved, or at least mitigated, by the use of smart cards as secure hardware modules. Smart card software is almost completely safe against tampering. This means that an attacker will find it very hard to access confidential content stored or processed on a well-designed system. In this way, the smart card can serve as a reliable outpost for the Service Provider, which continues to control it, even though it is permanently in the hands of an end-user.

Smart cards can also contain secret information, which is not available anywhere else, even on the manufacturer's or card issuer's site. A typical example is a signature key: such keys are generated on-card, are unknown to any outside entity, and always remain in the physical possession of the card holder. The card holder can therefore be certain that no one, not even the Service Provider or the card manufacturer, can forge his/her signature without physical access to his/her smart card.

Secure repositories realized on smart cards hence play a particularly important role in the CoSec security architecture. They do not only have the ability to securely store data but also to process security relevant protocols (e.g. issue and validate signatures, validate certificates, generate key pairs, etc.), providing functionalities that go far beyond secure storage.

Typically, execution of cryptographic protocols will require interactions between several of these. Due to the nature of the protocols and their underlying parameters, the interfaces between CoSec functions will inevitably be complex. Therefore, an in-depth description lies beyond the scope of this chapter; it rather aims to give the reader a full overview of CoSec and its capabilities.

6.2 Security Architecture

This section gives an overview of how security functions are provided to the Convergence system and are embedded into the system respectively.

From a methodological standpoint, the Convergence project defined security assets, considered threats to these assets, and derived security functional requirements, which drove the selection and development of appropriate cryptographic primitives, high-level functions, and protocols. This methodology is described in greater detail in (Anadiotis et al. 2011). In this section, only the main results of the security analysis are presented.

A key task of the methodology was to identify the "roles" that lie at the base of Convergence security. Note that this task is logically prior to the identification of the architectural entities (Technology Engines and Protocol Engines) that implement the required functionality in the Convergence architecture.

Two key roles are those of the Identity Provider (a trusted third party responsible for the registration of users, identification of users, verification of credentials, and issuing of certificates) and the Service Provider, the entity that handles "daily" business in a specific scenario, including user authentication and licenses. One of the benefits of separating the two roles is privacy protection: Identity Providers never gain access to data accumulated by Service Providers; Service Providers do not need to know all (or any) of the personal data that are collected during registration. A typical Identity Provider might be a government agency that performs end-user registration and supplies Service Providers (e.g. insurance companies) with users' certified credentials, but it could also be the operator of the Convergence platform.

As regards the derivation of security requirements from an analysis of assets and threats, we considered assets related to VDIs, assets related to users (end-users, peers, etc.) and services (service applications), and assets related to engines (software packages) in the Convergence middleware. A typical asset is for instance VDI integrity, referring to the VDI as a whole, including all of its components such as content, metadata, annotations, and the identifier: A VDI should never be changed, once it has been published, until possible revocation and the identifier must be unique for each VDI. Threats to VDI integrity include for instance the altering of components after publishing. This threat can be thwarted by cryptographic primitives such as signing a cryptographic hash of the VDI.

The system's main security features, identified during the analysis, are:

(i)   assurance of VDI integrity (and authenticity)
(ii)   governance of VDI access restrictions (confidentiality)
(iii) user identification and authentication
(iv)   issuing and enforcement of licenses
(v)   protection of user privacy
(vi)   network security.

Most of these features are provided at the middleware level (CoMid), some at the Computing Platform level. CoMid security features are provided by the Security TE, which in turn relies on features offered by CoSec in the Computing Platform. Some security features of the CoNet are provided by the CoNet itself; others require support from the CoSec.

The CoMid exploits the features offered by the CoSec component in the Computing Platform through a dedicated Technology Engine called Security TE. Components wishing to perform a security function or protocol use CoSec via the

Security TE. The CoSec serves all requests coming from the other engines and directed to the Security TE.

Based on CoSec, the Security TE can:

(i)   create new credentials and manage certificates
(ii)   generate keys and encrypt/decrypt data or keys
(iii) store confidential information (e.g. licenses and keys) in the secure repository
(iv)   certify the integrity of engines.

Other Engines rely on the Security Engine to perform:

(i)   signing of VDIs (VDI TE)
(ii)   symmetric encryption/decryption of resources (Media Framework TE)
(iii) symmetric encryption/decryption of keys (REL TE)
(iv)   user identification (Identify User TE)
(v)   user authentication (Authenticate User TE) (see Fig. 6.1).

Although the Convergence architecture diagram shows the CoSec component as a single monolithic block, it actually has a *distributed architecture* encompassing several independent (and possibly remote) components, each of which includes software as well as hardware.

CoSec components are distributed on the following computing platforms:

(i)   Client computers (e.g. end-user laptops);
(ii)   Application servers;
(iii) Peers (e.g. node computers in a network);
(iv)   Smart cards (typically held by the end-users).

Fig. 6.1 Security for convergence CoMid

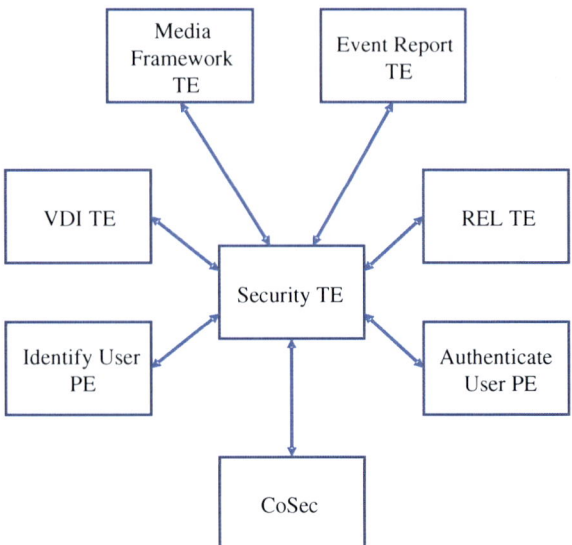

Most components of CoSec are located on *client computers* (e.g. end-user laptops), *smart cards*, *application servers* and *network* peers. The majority of protocols processed within CoSec involve several of these entities. Figure 6.2 illustrates the general style of these protocols as exemplified by a simplified user authentication protocol. In the example, the smart card acts as a secure repository for the user's signature private key.

CoSec features are exposed to the CoMid components (engines) through the CoMid Security TE. A CoSec API is defined so that the Security TE can use the CoSec.

In addition to the usage by CoMid through the Security TE, CoSec is also used by CoNet. CoNet supports security and privacy mechanisms aimed at preserving the integrity of the networking service and, where required, the anonymity of owners and consumers of named resources. A distinguishing aspect of CoNet security is the use of data-centric security: security information is embedded in CoNet data-units. Data-centric security makes it possible for user and network nodes to verify the validity of named-resources, preventing the caching and dissemination of fake versions. Protecting information at the source (i.e. protecting the data unit) is more flexible and robust than delegating this function to applications, or securing only communications channels. The basic security primitives required for these operations are provided by CoSec through the CoSec API.

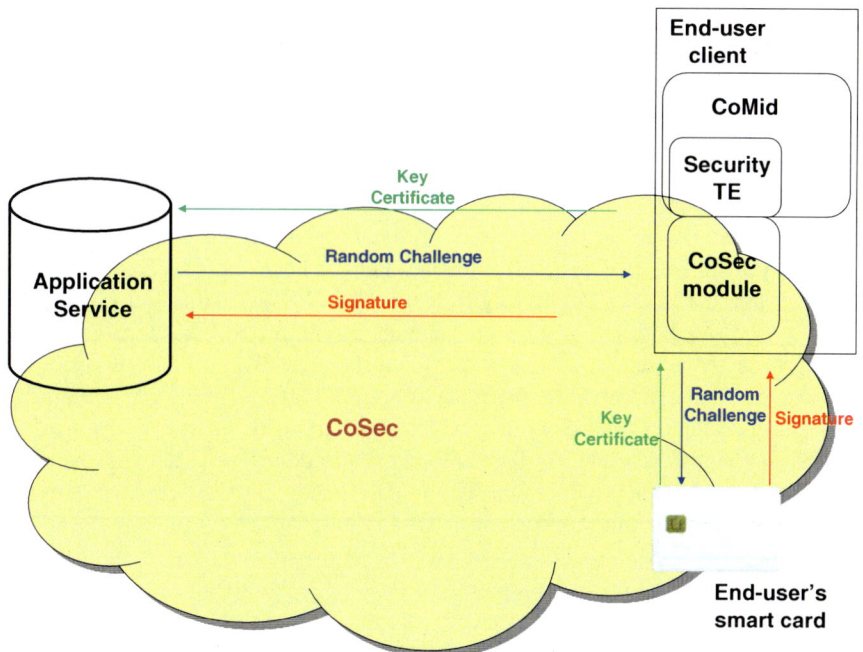

Fig. 6.2 CoSec elements and example information exchange

6.3 Cryptographic Primitives

To implement cryptographic functionalities, protocols or even applications, CoSec uses a couple of cryptographic primitives, which are described in this section. The primitives, which are implemented on smart cards to a large extent, are provided via the Security TE on the lowest level of the Convergence infrastructure and thereby exposed to the upper-level Convergence components. Thus, middleware components or applications can directly access them in form of stand-alone crypto-functions. In one of the following sections, we are going to describe the implementation of cryptographic protocols based on the primitives described here. These protocol implementations, secured by execution on smart cards, are also part of CoSec.

In order to ensure confidentiality, fast symmetric encryption and decryption methods for Convergence content VDIs are available through an implementation of the common AES scheme. In order to enable seamless integration of the Convergence secure token into the overall architecture, we selected the CBC chaining as it is the common mode for smart cards. The Security TE currently supports AES keys with up to 256 bits. They can be chosen dynamically and configured according to the requirements of the actual Convergence application or protocol. In addition to AES, the Security TE provides ABE (Attribute-Based Encryption) (Bethencourt et al. 2007; Lewko and Waters 2011) and IBE (Identity-Based Encryption) (Galindo et al. 2009; Martin 2008) symmetrical encryption primitives used for key derivation inside Convergence.

As regards asymmetric cryptography, we first of all support RSA with a key length of up to 2048 bits. Access to private keys is encapsulated in functions implemented on the smart card based Convergence token. This is possible as the smart cards' crypto-processor is especially designed for efficiently computing complex RSA operations. Asymmetric cryptography is most notably used for key agreement, signature generation and verifications, as well as for the support of certificate-based security within CoSec protocols. Thus, on top-level of CoNet it is the main primitive used for authentication and the licensing of VDIs.

A system based on a Public Key Infrastructure (PKI) (Adams and Lloyd 1999) utilizes the advantages of certificates generated by a trusted third party in order to establish trust within a system. The Convergence PKI will be described in Sect. 6.5.4. The support for a structure as complex as PKI, requires a number of basic cryptographic functionalities, which are part of Security TE. Each of the functionalities is embedded in the Convergence middleware to support the whole set of possible identities, components and data inside the Convergence network.

Basic primitives such as hashing (SHA-256, SHA-2, etc.) or random number generation are also part of the Security TE. They are required in almost every cryptographic operation, e.g. in mutual authentication protocols, basic encryptions schemes, as well as in the advanced encryption schemes implemented within CoSec.

6.4 Advanced Cryptographic Schemes

As described in the previous section, we do not only use standard cryptographic primitives, but our architecture also supports advanced schemes for anonymous group signature and Identity-based encryption, both based on Pairing based cryptography. Moreover, we support pseudonymous access by "Restricted Identification". These schemes and their application in Convergence are described in this section.

6.4.1 Group Signature Schemes

Group Signature Schemes (Boneh et al. 2004) allow any member of a previously created group to digitally sign a document or an object (in the case of Convergence a VDI or a part of a VDI) on behalf of the entire group. Here, the term 'group' refers to a set of people that perform equivalent tasks and do not necessary need each other's consent (e.g. editors of Wikipedia). The group has a single public key that a third party can use to verify if the signature originates from a member of the group. However third parties cannot ascertain the identity of the group member responsible for the signature. If the same member of the group signs on multiple occasions, the signatures cannot be linked (Libert and Vergnaud 2009). In other words, group signature schemes do not just provide pseudonymity, but also anonymity (Calandriello et al. 2007), except in rare conditions, when the group master can legitimately unveil the identity of a user.

Group Signatures are quite a recent development and legal implications/ requirements related to their use have yet to be defined. In general, a group is administrated by a trustworthy group master, who holds a master key with which he can register new members, revoke existing members, and unveil the identity of a signer in the event of a dispute. This ability may be a procedural requirement in some applications, where unconditional anonymity is not permitted.

Group signatures are employed in Convergence for basically two scenarios:

(i) *Anonymous authentication on behalf of a group*: Access to services can be granted based upon an SSO (Single Sign-On) protocol, for which in turn a smart card based initial authentication is performed. For specific applications, such an authentication can be an anonymous group authentication based on a group signature scheme.

(ii) *Anonymous signing*: In contrast to (i), where a random token is signed for authentication purposes, here content is signed. However, due to the nature of the content, it is sometimes desirable to sign anonymously. This allows for data protection and privacy, while at the same time it admits to reveal the identity of a member in case of a misuse. Such a revealing process can only be executed under very strict regulations however. Convergence employs this feature in the LMU scenario, as outlined in the subsequent paragraph.

The LMU scenario provides an example of the way a group signature scheme could be applied. In this scenario the group consists of all registered students who have access to at least some of the lectures offered, and all lecturers. Membership requires a registration process. During this process, each student receives a smart card that includes both a personal signature key (used to issue personal digital signatures), and a private group signature key that enables the student to generate an anonymous group signature, proving that he or she is indeed a valid member of the corresponding group of students. For instance, a student, who proves she is part of a group attending a course, can sign annotations to course materials.

6.4.2 Identity Based Encryption: Attribute Based Encryption

Recent years have seen significant progress in so-called Identity Based Encryption (IBE) (Galindo et al. 2009; Martin 2008) and Attribute Based Encryption (ABE) (Bethencourt et al. 2007; Lewko and Waters 2011). Cryptographic protocols based on these schemes have now matured to a point in which they can be deployed in real life applications.

IBE provides asymmetric cryptography (encryption and signatures) without the burden of certificate administration typically associated with the creation of a PKI. By using IBE, Convergence can offer encryption and decryption of VDIs without establishing a cumbersome PKI (Adams and Lloyd 1999).

This feature is of particular interest in cloud computing scenarios. Before encrypting data and uploading them into a cloud, a sender must be sure about the identity of the recipient's public key. Since files in the cloud are easily physically available, and no connection-based session encryption can be guaranteed, it would constitute a major security breach to encrypt a data package to an unintended recipient on the basis of a non-trustworthy public key. Conventional asymmetric cryptography ensures security through certificates, at the cost of the burden coming along with certificate administration.

IBE schemes also offer the possibility to generate digital signatures without having to deal with certificate administration (Adams and Lloyd 1999). This is employed in CoNet to authenticate the origin of data packages at each node prior to processing them. Note the importance of checking the authenticity of packages during transport and not merely at the recipient's site in order to thwart Denial-of-Service attacks.

ABE makes it possible to map complex licenses to attributes. The basic idea of an Attribute Based Encryption scheme is that content (or more precisely a content key) is encrypted according to determined attributes. In this way, only key holders with the attributes required by the encrypted content will be able to decrypt it. This avoids the risk that fraudulent users could bypass licensing by tampering with security-related hardware and software: while a license check may be bypassed (for instance through manipulation of proprietary hardware like an e-book-reader), recipients of ABE-encrypted content who do not have specified attributes (e.g. are under 18 according to the personal data on their smart card) will be unable to decrypt it.

6.4.3 Pseudonymous Access: "Restricted Identification"

Within the context of this report, we use Pseudonymous Access as a synonym for the specific "Restricted Identification" technology, as it is used for instance for the German Electronic Passport. This is a technology that allows a registered user to derive a pseudo-identity from a personal private key contained in his/her smart card, and which, where necessary, allows administrators to revoke or black-list these identifiers. The same technology can also be used to derive "sector"-specific pseudonyms, where a sector could be a geographical area, a field of application, a set of transactions with a particular provider, or many other things. This means that, in a given sector, an individual user will always use the same pseudonym but that in other sectors she will have different pseudonyms, which cannot be linked to each other.

As an example, the Convergence smart retailing scenario can use Pseudonymous Access to introduce data protection and improve privacy for users of the system. In this scenario, customers use their shop specific pseudonyms to order goods without revealing their true identity. Pseudonymous Access prevents shops from cross-linking their data and prevents third parties from mining data from different shops to create profiles of individual customers.

6.5 High Level Security Support

The previous sections have described the security architecture, the available cryptographic primitives and the advanced cryptographic schemes for CoSec. In addition to these, CoSec includes a number of high-level functionalities to ensure authenticity and integrity of identities, users, components, and VDIs within the Convergence network. The methods explained in this section combine the described functionalities of the previous sections into different higher level security protocols or structures which are also available to the Convergence network through CoSec. Here, the Security TE, orchestrator engines, protocol engines and elementary services of Convergence form the architectural pattern on which the high-level functionalities are based on. Applications running inside Convergence are able to use them to securely connect to and communicate within Convergence.

Authenticity of identities, such as users, servers, etc., inside CoNet is achieved through authentication protocols, e.g. challenge-response protocols involving signatures or key-agreement protocols that do not involve end-user signatures.

Another critical point is the integrity and authenticity of a VDI. A VDI must remain unaltered throughout its lifetime, i.e. after publishing and before revocation. Where required or granted by the creator, the origin of a VDI must remain unchanged throughout its lifetime. In such cases, a user shall be able to gain assurance about the authenticity of a VDI. Depending on the VDIs' policies, users may be allowed to create VDIs anonymously, or on behalf of a group of users, or under a pseudonym.

Applying policies on VDIs requires licensing which mainly involves the issuing of licenses and license enforcement. To enforce policies, the Rights Expression Language Technology Engine (REL TE) heavily relies on card-based security functionalities also explained in the following section. An on-card license validation, using card verifiable certificates and an Attribute-Based Encryption, mapping license conditions to attributes are involved here.

(1) User Registration

In the Convergence user scenarios, an Identity Provider administers registrations. The registration process involves two steps: identification and cryptographic key generation. On registration, each end-user receives a smart card containing at least the key sets described in the following along with a unique identifier. Each user can retrieve this identifier based on the key sets of his/her smart-card.

- A unique signature key pair
  The key is generated on the smart card itself, during registration, according to distinct parameters. The secret part of the signature key is stored in such a way that it can never be read out of the smart card by any entity whatsoever.
  The key is certified during registration. The certificate (containing the public key part) is signed by the registration authority and stored on the smart card, as well as on a Convergence server. The generation of this key pair is mandatory for any user wishing to use a smart card as secure repository.
- A unique encryption key pair
  Each user is assigned a private–public key pair, generated according to the same policy chosen for signature key generation. The encryption key pair is different from the signature key pair: using the same key pair twice is prohibited; the processes used for generating the two key pairs must be entirely independent.
  The key is again certified during registration; and the certificate is stored on the member's smart card, as well as on Convergence servers. The generation of this key pair is mandatory for each user wishing to use a smart card as a secure repository.
- A unique group signature key
  During registration, each member's smart card generates a random seed, securely transmitted to the group administration authority for computation of the complementary secret key part. The user's signature includes both parts, both of which must be safely stored on his or her smart card. Note that the group

signature key, unlike the signature key and encryption key, is known not just to the smart card but also to the "group master" (see paragraph 6.4.1.)

- A unique private key for Identity/Attribute Based Encryption

Like the group signature key discussed in the previous paragraph, the private keys for users of Identity Based Encryption and Attribute Based Encryption are derived from a "master" key held by a trusted authority. The setup is similar to the setup for the group signature scheme.

In Convergence the "Identify User" elementary service uses the "Identify User" protocol engine for user registration. This protocol engine provides the identify user method from the "Identify User" engine orchestrator. The method receives an MPEG data object containing user details and the certificate. Upon successful registration, a unique user id is returned wrapped in an MPEG response containing the user id and additional user information.

(2) Authentication

When an end-user wishes to authenticate him/her to an application, he/she executes the following procedure:

(i)   The client presents his/her user certificate for the current application (obtained from a Service Provider) to the Application Server.

(ii)   The Application Server sends its own certificate (issued to it by an Identity Provider) to the client.

(iii) The client validates the Application Server's certificate.

(iv) The server performs server authentication if required by the client (including certificate validation by the client).

(v)   The Application Server validates the user certificate.

- Validating the issuing Service Provider's signature and checking for attributes encoding access rights, privileges, etc.
- Validating (if not already done) the Service Provider's own certificate (with the public key from the Identity Provider's root certificate), obtained from the SP's database, or from the end-user's smart card.

(vi) The Application Server requests a client authentication based on challenge-response.

(vii) Both parties continue with secure messaging, if required by one of the participants.

If an application has previously validated the certificate of a Service Provider and has labelled its public key as trustworthy, Step (v) may not be required. Additionally, it may also be admissible or even required—for an end-user to authenticate under her "real" identity as issued to her by an Identity Provider. Finally, the public key of Identity Providers is considered to be trustworthy, i.e. we assume that they are contained in root certificates, which inherit their trust through other channels and not through yet another certificate.

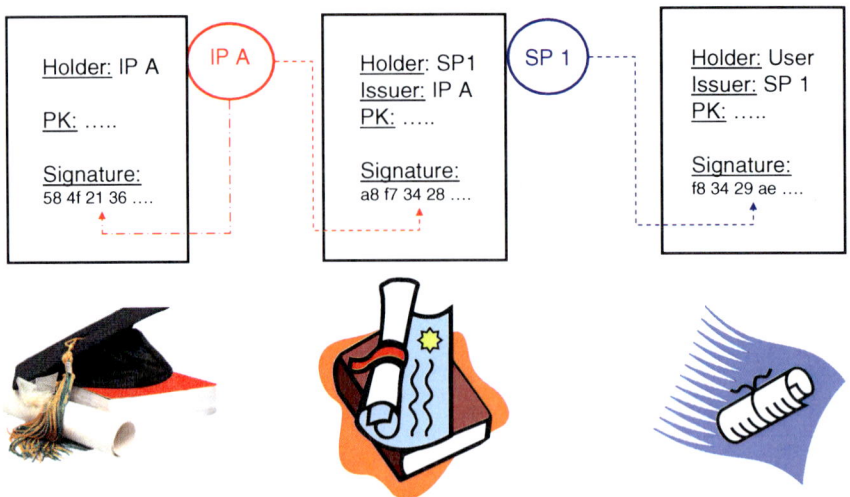

Fig. 6.3 Certificate chain

Regarding certificate chains, Fig. 6.3 illustrates a simple certificate chain consisting of three certificates:

(i)   The (self-signed) root-certificate of Identity Provider IP A;
(ii)   The Service Provider SP 1's certificate, signed by IP A;
(iii) An end-user's certificate, signed by SP 1.

Note that the diagram does not show the "real-name" certificate that the end-user has received from IP A and has presented to SP1 to obtain the right-hand side pseudonym certificate. Validation of this certificate is not part of the chain.

The full validation chain is composed of three steps:

(i)   Validation of the root-certificate (integrity check with attached signature);
(ii)   Validation of SP 1's certificate (with public key extracted from IP A's certificate);
(iii) Validation of User's certificate (with public key extracted from SP 1's certificate).

Provided that a verifier (e.g. a Convergence application) trusts the IP A to properly execute its duty, these steps will be sufficient to establish trust in the user's public key and accompanying attributes without previously having established trust with any Service Provider. Although this chain looks artificially simple, real-world certificate chains are often not much longer. Chains with three hierarchy levels are common.

(3) Signature issuance and verification

As can be seen from the authentication scenario, each end-user, Service Provider and Application Server always has one or more signature key pairs.

Service Providers and Application Servers only sign with key pairs certified under their real identity. Corresponding certificates are issued to them by Identity Providers. End-users may sign on behalf of their registered pseudonyms. The corresponding certificates are issued by Service Providers. They may also sign using their real identity, again requesting the corresponding certificate from a Service Provider. Verifying a signature involves certificate validation, and works in the same way as authentication. Only after successful certificate validation the verifier of a digital signature can be sure that a signature has really been issued by the entity claimed. Each verifier of a signature is responsible for establishing this trust.

(4) Encryption and decryption

Since it is impossible to use the private key embedded in the smart card for decryption or for purposes such as signature and authentication, Encryption and decryption with asymmetric key pairs requires two keys. Service Providers and Application Servers can obtain certificates from Identity Providers. (An Identity Provider provides the signed assertion of user to a Service Provider). End users can obtain certificates for their real names and pseudonyms from Service Providers.

An entity wishing to encrypt a message to a public key of another entity has to validate the designated recipient's corresponding certificate prior to using its public key for encryption. This means they need to obtain the appropriate certificates from server databases, or obtain them via mail or other means. Certificate validation involves the same kind of chain validation previously described for the authentication procedure.

Typically, an end user has to:

(i)   Validate the recipient's certificate;
(ii) Validate the certificate of the Service Provider who issued the recipient's certificate.

Validation of the recipient's certificate before encryption to the associated public key is essential. It ensures that a message is encrypted for the designated recipient and thereby thwarts "man-in-the-middle-attacks".

6.5.1 Key Generation of Asymmetric Key Pairs/Certificate Storage

A certificate must be accompanied by an asymmetric key pair, containing the public key of the holder. The details differ depending on for whom the certificate was issued.

Table 6.1 Identify provider key pairs

Identity provider	Purpose
"master" key pair	Signing/issuing certificates
[optional further key pairs]	Authentication/signing/decryption

Table 6.2 Service providers key pairs

Service provider	Purpose
"master" key pair	Signing/issuing certificates
[optional further key pairs]	Authentication/signing/decryption

- Identity Providers
  Identity Providers hold an asymmetric key pair they use to sign certificates they issue (certificates issued on behalf of end users, Service Providers and Application Servers). It is referred to as "master" key pair in Table 6.1.
  An Identity Provider's certificate-signing key pair is considered to be its "master key". Compromising this key would endanger the entire system of trust depending on it. The private part of the key is thus the Identity Provider's "sanctuary" and needs to be stored in a highly secure way. The public part is published in the root certificate of the Identity Provider.
  As shown in Table 6.1, Identity Providers may also have additional key pairs for "ordinary" tasks such as authentication, signing (of VDIs, messages, not certificates), and decryption.
  Identity Providers have to maintain a database containing all certificates they have issued, namely:

  (i)   End-users' "real name" certificates;
  (ii) Service Providers' certificates;
  (iii) Application Providers' certificates.
        This database is needed for purposes such as revocation, renewal, duplicate enrolment checks, etc.

- Service Providers
  Like Identity Providers, Service Providers hold an asymmetric key pair for signing certificates they issue to end-users, as shown in Table 6.2.
  Service Providers have to provide a database containing all certificates they have issued, namely:

  (i) End-users' "pseudonym" certificates, along with an entry of the end-users' real identities;
  (ii) Application Servers' certificates (optional).
       The Service Provider uses the database for purposes such as revocation, renewal, duplicate enrolment checks (in cases where duplicate enrolment is prohibited), and tracing the real identities behind pseudonyms (when requested by an authorized party). Note that the database must not be

Table 6.3 Application servers key pairs

Application server	Purpose
Authentication key pair	Server authentication
Signature key pair	VDI signatures
Encryption/Decryption key pair	Decryption

publicly available; otherwise it would be possible to link pseudonyms to the real identities behind them. Only Service Providers should be allowed to do so, not even Identity Providers!

- Application Servers
  Application Servers hold asymmetric key pairs in very much the same spirit as end-users (cf. Table 6.3).
  In Convergence, it is a requirement that a client connecting to an application should be able to authenticate the server to which she is connecting. Convergence therefore requires that all Application Servers should hold a certified authentication key pair. The corresponding certificates may be issued by Identity Providers, or optionally by Service Providers. Note that Application Servers need to be identified under their "real" names.
- End users
  End users hold asymmetric key pairs for tasks like authentication, digital signing and encryption/decryption (apart from their key pair(s) originally obtained during registration by an Identity Provider). The private part of these key pairs is generated and permanently stored on the end user's smart card. These smart cards are secured in such a way that a private key part cannot be output and will therefore never leave the card. The end user's card stores the following key pairs for each Service Provider it uses. In the example depicted in Table 6.4, there are two Service Providers. The first (SP1) offers a service through the application Appl1, the other (SP2) through Appl2. Note that some fields may be empty (for instance, if encryption/decryption is not required for certain applications).
  Please note that there is only one entry for an end user's "real-name" key pair, and that this key pair is used for authentication purposes only. Its main use is to authenticate the user to Service Providers when requesting certificates from Service Providers' for his (the user's) own key pairs.

6.5.2 Security and Privacy in Key Generation Procedures

There are a number of concerns which have to be taken into account when keys are generated, regarding security and privacy aspects.

- Security
  All asymmetric key pairs associated with Identity Providers, Service Providers, Applications Servers, and End-Users shall be generated by the entities owning them.

Table 6.4 End users key pairs

End user	"real-name" key pair	SP 1	SP 2
Authentication key pair	Authentication towards SPs	Authentication before Appl 1	Authentication before Appl 2
Signature key pair		VDI signature issuance	VDI signature issuance
Encryption/ decryption key pair		Decryption of received messages/VDIs	Decryption of received messages/VDIs

The corresponding private key parts are kept by their respective owners and are never delivered to any other user. In particular, end users' private keys are generated and stored only on their smart cards, and are never an output of any function. This implies that not even the end-user herself will ever know her own private key.

- Privacy

  The fact that a party issues a certificate for a key pair does not, and indeed must not, imply that the same party issues the key pair. Parties issuing certificates should verify ownership of private keys using cryptographic protocols such as challenge-response or key agreement. Thus Identity Providers, who occupy the highest level in the Convergence trust hierarchy, do not know any private key except the keys they directly own. This means that they are not able to forge signatures on behalf of any entity they certify. This is an important privacy issue. Although end-users need certificates issued by Identity and Service Providers, they need to be certain that these providers cannot sign messages on their behalf, or decrypt messages encrypted to their public keys, or authenticate under their name or pseudonyms.

- Key pair generation

  Key pair generation follows established standards, and meets current security requirements with respect to choice of algorithm, key length, parameters, etc. Smart card issuers are responsible for securing cards in such a way that they respect regulatory requirements for key-length, key generation algorithms, etc. The card issuer can also enforce additional conditions for key pair generation. For example, it can bind the generation of specific key pairs to a role (to be verified by role authentication) so that only legitimate terminals can activate key generation. In Convergence, a smart card can generate RSA keys of bit-length 1024 or 2056. When a key pair is generated, basically three types of key pairs are generated, which are named 'sig' for signature generation and verification, 'auth' for the encryption/decryption in challenge/response, 'conf' as the key for data encryption and decryption. Card based asymmetric encryption is the most secure as the private key of the card cannot be retrieved.

  Convergence assumes that key generation by Service Providers and Application Servers complies with established policies. In a real-world scenario compliance may involve auditing by certified labs or governmental organizations.

6.5.3 Use of Keys, Owner Authorization

End-users hold their private keys on their smart cards, and do not have access to these keys. Strictly speaking, this implies that it is not the user who performs authentication signing or decryption, but the smart card. In order to safeguard against loss or misuse of smart cards, private key operations require authentication of the card holder, before they are performed.

This authentication is usually performed using well-known methods like:

• PIN authentication or
• Biometric authentication with on-card verification.

Convergence smart cards use PIN authentication. A retry counter limits false PIN entries to a reasonable number. Other entities like Identity Providers, Service Providers, Application Servers, etc., which are not using smart cards for their private keys, have to store private keys in a secure repository allowing a meaningful authorization (e.g. by passport, hardware token, etc.).

6.5.4 Public Key Infrastructure

Convergence security is constructed on a PKI architecture in which certificates provide a standard means to inherit trust. As described above, in certain applications, the PKI can be replaced with IBE.

Very basically, certificates help to perform two basic functions.

• They bind a public key to an entity (human user, device, network node, institution, etc.) in a trusted way;
• They allow a trustworthy authority to associate attributes with entities.

Certificates are issued by Central Authorities (CAs), whose defining property is that users trust them. A CA is a role that an entity can assume based on specific and agreed policies for a specified group of users. Thus a CA may be a governmental organization responsible for the citizens of its country; or a manufacturer acting on be-half of a group of retailers and customers, or even an individual user who acts as a CA for a group of co-users (usually a small group).

In Convergence there are two kinds of Certificate Authorities (CAs):

• Identity Providers;
• Service Providers.

Convergence certificates comply with the X.509 standard. Each certificate is associated with:

(i)   A holder: the subject (=owner) for whom the certificate has been issued;
(ii) An issuer, which is the entity that issued (most importantly: signed) the certificate for the owner;

(iii)   The holder's public key (accompanied by attributes referring to the associated algorithms used);

(iv)   The issuer's signature (accompanied by information referring to the corresponding algorithm used for signing);

(v)   Additional attributes, referring to the holder and/or issuer.

(End-) Users are registered once by an Identity Provider, under their "real identity", and then by various Service Providers for different applications, possibly under different pseudonyms.

- Registration of an End-User by an Identity Provider
  When registering to a Convergence Identity Provider that he/she completely trusts, a user (an end-user, a Service Provider, an Application) has to reveal his/her real individual identity (name, address, passport data). The Identity Provider issues a unique user identifier on behalf of the user, supervise asymmetric key pair generation, issue smart cards, and issue a certificate. Each user may only register once ("duplicate enrolment" is forbidden). The Identity Provider is responsible for verifying a users' claimed identity (e.g. by checking his/her passport), as well as for revocation of certificates, and duplicate enrolment checks.
  A "real world" infrastructure would require complex certificate chains and procedures to verify Identity Providers' public keys (e.g. audited personalization during smart card manufacturing, publication of keys over trustworthy public channels, personal handover, etc.). In the trials, we assume, for the sake of simplicity, that these keys are certified by self-certifying root certificates.
- Registration of an End-User by a Service Provider
  Convergence Service Providers are responsible for assigning specific user identities for use with specific applications. Such identities may be pseudonyms that hide a user's real identity from the application, or real identities. Service Providers' public keys are certified by a Convergence Identity Provider; the same holds for the public keys of Application Servers.
  When users register with a Service Provider they do not have to use the same pseudonym they have used with other Service Providers. This avoids cross-linking. Detailed conditions may be subject to policies in place for specific Service Providers and Applications. End-users can also apply to Service Providers for certificates on key pairs they use for signing or for decryption/encryption. Identity Providers only issue certificates for authentication key pairs. Each Convergence application cooperates with one or more Convergence Service Providers it trusts (e.g. with whom it holds contractual agreements). Service Providers may cooperate with more than one application, and applications may cooperate with more than one Service Provider.

The design discussed above means that the Convergence demonstrator architecture supports three levels of certificates:

(i)   (Self-certified) Root certificates for each Convergence Identity Provider;
(ii)   Convergence Service Provider and Application certificates issued by an Identity Provider;
(iii) End-User certificates issued by Identity Providers, or Service Providers.

Identity Providers are trustworthy "by definition" (i.e. by assuming an established process ensuring their trust). Their root-certificates are self-signed and contain the public counterpart of the corresponding certificate-signing key.

Service Providers are certified by an Identity Provider. The certificates issued by the Service Provider itself use its own certificate-signing key and are used to prove to application servers that the pseudonymous certificates it has issued to end-users do in fact come from a trustworthy Service Provider. Service Providers may also hold a certificate on its authentication key pair. Identity Providers also certify the authentication key pairs for Application Servers and (optionally) their signing key pairs.

Figure 6.4 depicts the basic infrastructure. In the diagram, fat arrows show who issues a given certificate. Thus an end-user obtains her (genuine) certificate from an Identity Provider (IP). She then presents this certificate (containing her real personal name and data) to two different Service Providers (SPs) that she trusts and requests a pseudonymous certificate from each of them. In this way, the end-user can apply for certificates on each of her key pairs for each of the three categories described above, i.e. for authentication, signature issuance and decryption.

Example

The following table illustrates the content of an end user's smart card when she interacts with three different applications. To begin with, she generates a single key pair that is certified by an Identity Provider—probably with a procedure that requires her to appear in person. The key pair is stored on her smart card and is labelled with the name of the Identity Provider (IP) (see the second column in Table 6.5). Using this certificate she uses online channels to contact various Service Providers, who provide her with certificates for the following additional key pairs she has generated on-card

(i)   For Application 1, she requests certificates over key pairs for authentication, signing and encryption from Service Provider 1 (see column 3, Table 6.5);
(ii)   For Application 2, she needs only key pairs for authentication and signing, and requests certificates on them from Service Provider 2 (see column 4, Table 6.5);
(iii) For Application 3, she needs only a key pair for signing, and requests a corresponding certificate from Service Provider 3 (see column 5, Table 6.5).

She stores the certificates on her smart-card. Optionally, the Service Providers may distribute these certificates to designated databases. Note that the trust in certificates does not stem from the place they are stored or the entities that (physically) dispose of them, but from the issuer's signature.

Remark: In this way, end-users' smart cards can serve as decentralized databases for certificates. It is very convenient for an authenticating server to obtain

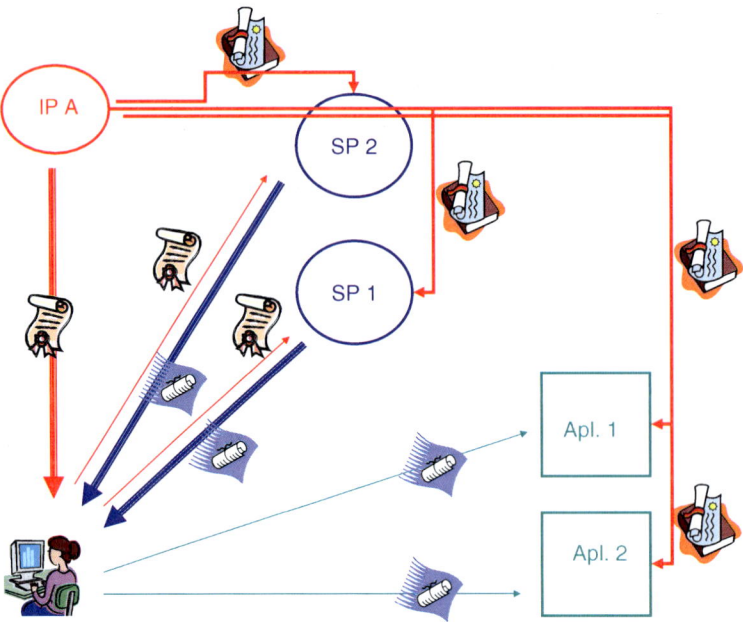

Fig. 6.4 Certificate infrastructure

Table 6.5 Key-pair and Certificate

	IP	SP1 Appl. 1	SP2 Appl. 2	SP3 Appl. 3
Authentication	Key pair + certificate Key_UIP	Key pair + certificate Key_Auth_1	Key pair + certificate Key_Auth_2	
Signature		Key pair + certificate Key_Sig_1	Key pair + certificate Key_Sig_2	Key pair + certificate Key_Sig_3
Encryption/ decryption		Key pair + certificate Key_Encr_1		

the necessary certificates directly from the requesting end-user, instead of having to store them all or requesting them from a remote database in real time.

Note that it may be admissible that one and the same Service Provider generates certificates for different applications. In terms of the example below, SP1 = SP2 although Appl. 1 differs from Appl. 2.

Some applications may request sensitive data stored on a smart card (e.g. a stored fingerprint template, medical records), try to store something on the card (e.g. new certificates), or initiate sensitive on-card operations (like key generation). In such cases it is the card itself that ensures that the request comes from a legitimate source, even against the will of the card holder. For instance, a card issuer might encode a health card in a way that only authorized physicians have a right to download medical records. The card will refuse to reveal such records to non-authorized physicians, even if the card holder (the patient) is willing to allow this.

To this end, certificate validation and authentication token validation are performed on-card. This scenario is called "Role Authentication". A special (compact) format for Card-Verifiable (CV) certificates supports efficient certificate processing. As in the off-card scenario, the smart card needs to verify a certificate chain. Root certificates have been stored on-card during personalization by a trustworthy process, and often contain only a root public key rather than a complete certificate. The integrity of such root public keys is ensured by special mechanisms (e.g. writing into secured memory areas, check-sums, etc.).

In order to avoid a cumbersome structure involving blacklisted certificates, and to ease revocation and renewal, certificates have a limited period of validity. The validity period is part of the certificate, and its proper validation is naturally part of each certificate evaluation. Each certificate carries at least two dates: effective date of generation/issuance, and the expiration date.

The normal procedure is to periodically renew all certificates that have not been revoked. The validity periods for existing (expiring) and renewed (replacing the expired) certificates overlap, allowing enough time for distribution. Typically, within a hierarchical PKI, validity periods are shorter for entities further down in the hierarchy.

6.6 Single-Sign-On User Authentication via Third Party Identity Provider

For the Convergence trials, an authentication scheme has been implemented which features Single Sign-On (SSO) combined with a sophisticated user authentication through smart cards. In this scheme, the task of authentication is assigned to a trusted third party that verifies the user's identity, based on the credentials provided during registration, and provides the user with a token that the application uses to decide whether or not to grant him/her access, and to define the access privileges. The Authenticate User protocol uses Security Assertion Markup Language (SAML) to perform the user authentication. In Convergence, an authentication provider uses the Authenticate User elementary service in the middleware to perform the authentication. The authentication (or SSO) provider can thus be seen as an application, used by other applications to perform the authentication in their place. When a user tries to access a restricted area of the application, the application redirects the user to the SSO provider, which performs the authentication. The authentication result is a SAML assertion, passed to the application. The application then decides whether the user will get access to the requested area.

In the following, the procedures for registration and authentication are described. They have been adapted from MPEG-M.

- User registration
  User registration is a common procedure in almost all modern applications, especially applications that exploit user preferences to select their content. During registration, users provide the personal information needed by the application. This personal information is then linked back to them using an authentication procedure. The result is the assignment of a unique identifier to the user that distinguishes him or her from other users with different preferences. This identifier may be the e-mail address, a name defined by the user or a name generated by the application.
  Convergence has specified a user identification protocol, which takes as input specific data associated with the user to be identified. This data could be a fingerprint scan, an iris scan, a digital certificate, a password, etc. The choice of what data to use depends in some cases on the user, in others on the applications. On success, this protocol returns a user identifier, chosen in line with the policy of the "Identify User" Service Provider (SP). The user identifier may be a random string, the user's e-mail, etc.
  One key issue here is the correctness of the personal data provided as input to the "Identify User" SP. For example, a user might use a false fingerprint to prove his or her identity. At this point, it should be clear that the Identify User SP is not a registration authority in the Public Key Infrastructure (PKI) sense and that it is still possible to use a PKI or a similar structure to ensure that the user is indeed who he or she claims to be. Thus, when the user provides a fingerprint, the "Identify User" SP may require that this fingerprint is signed by a trusted authority (e.g. the user's national police department). In some cases, a trusted third party can provide the Identify User SP with a SAML (Security Assertion Markup Language) assertion, guaranteeing for the user in advance of the identifier being issued by the "Identify User" Service Provider. In this way, the third party acts as an authority trusted by both the application and the user, making it possible to "hide" the real identity of the user behind an "anonymous" identifier, while still allowing the user to perform critical operations, such as money transactions.
- User authentication
  The goal of the user authentication procedure is to confirm that the user is who he or she claims to be. It is thus closely linked to the registration procedure. The authentication scheme we describe here is based on the SSO principle. In this scheme, the task of authentication is assigned to a trusted third party that verifies the user's identity, based on the credentials provided during registration, and provides the user with a token that the application uses to decide whether or not to grant him/her access, and to define the access privileges. The Authenticate User protocol uses SAML to perform the user authentication. In Convergence, an authentication provider uses the Authenticate User elementary service in the middleware to perform the authentication. The authentication (or SSO) provider

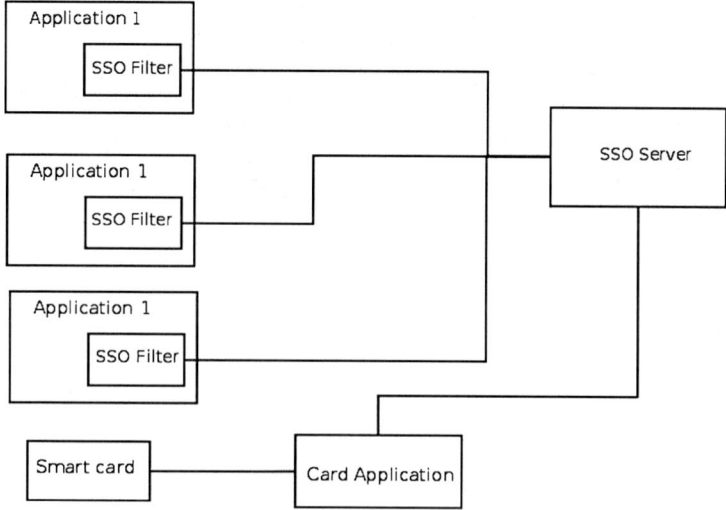

Fig. 6.5 Component diagram of the SSO architecture

can thus be seen as an application, used by other applications to perform the authentication in their place. When a user tries to access a restricted area of the application, the application redirects the user to the SSO provider, which performs the authentication. The authentication result is a SAML assertion, passed to the application. The application then decides whether the user will get access to the requested area.

6.6.1 Authentication with SSO

SSO authentication requires communication among various software components as it is depicted in Fig. 6.5. Each application (e.g. Application 1) interacts with the SSO server through a dedicated component called SSO Filter. It handles the SSO logic (login and logout) and accordingly sets sessions and does proper redirect between the SSO server and the application.

The actual process of an authentication based on the SSO principle is shown in Fig. 6.6 as an Interaction Diagram.

A user can register in the SSO server either with a smart card or via user name and password based registration. For card-based registration, the user requires his or her Convergence smart card and the native card application the interface of which is depicted in Fig. 6.7. SSO server also allows normal cardless registration as shown in Fig. 6.8. During both kind of registrations, he/she is also asked to provide an alias (can be used used as username) and a password (which would not have been necessary in principle using a card-based registration), so that both card-based as well as username/password based authentication can be used later on.

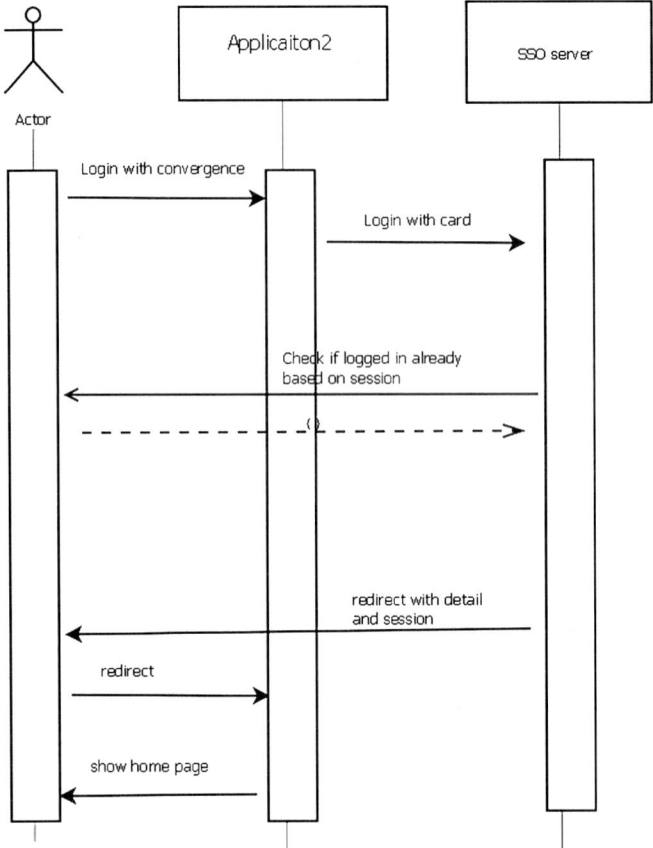

Fig. 6.6 Interaction diagram of SSO

Once a user has been registered, there is the need to authenticate the user for later activities. Using a card-based procedure, the authentication works as follows:

(i)   User clicks login via convergence (Fig. 6.9) then user is redirected to authorization end point as shown in Fig. 6.10. A user enters his alias name, and inserts his smart card into a card reader. He has to authenticate against the smart card, usually with a PIN or a password.

(ii) Then a mutual challenge response protocol is initiated where both server and smart card verify each other's identity. Trust relationship between the smart card and the server is established through:
      • Validation of corresponding certificates and
      • The mutual challenge-response protocol based on digital signatures.

(iii) Upon a successful two-sided validation the user is automatically logged in as described in section "Logging into a different application".

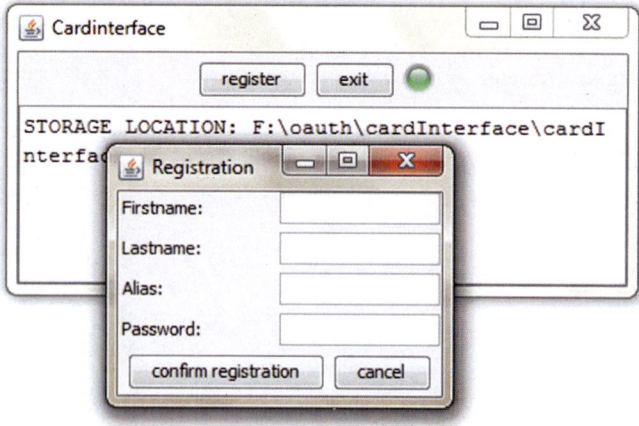

Fig. 6.7 Registration form for card application

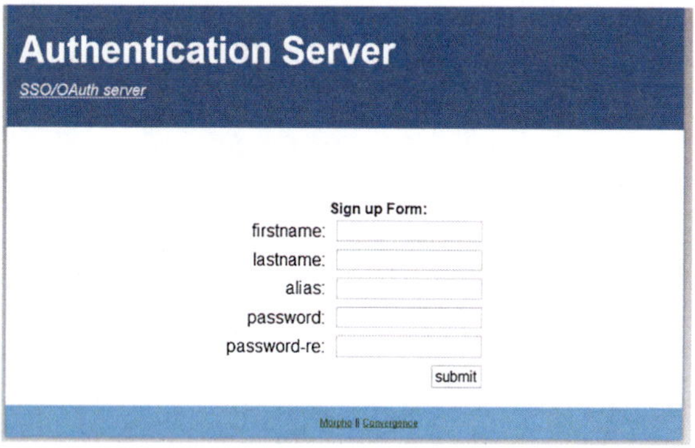

Authentication Server
SSO/OAuth server

Sign up Form:

firstname:

lastname:

alias:

password:

password-re:

submit

Morpho II Convergence

Fig. 6.8 Registration form for a web-based application

In the case where no card is used, the authentication procedure is based on a user name and password where the user provides his alias name and password and gets logged in, interface for this is shown in Fig. 6.11. In this approach the trust is established based on user's browser verifying the certificate or user him-self verifies the certificate. This approach is however known to be much less secure than the smart card based one. It is for instance prone to phishing (Crowd 2012). After server validation of the password, the user is logged in as described below.

(i)   When any request comes to an application (e.g. to the Alinari demo application as depicted in Fig. 6.9) the request is first intercepted by the SSO filter.

(ii) The filter then redirects the client's browser to the SSO server redirect end point, either to the card based SSO server as depicted in Fig. 6.10 or—in case no smart card is detected—to the cardless server as depicted in Fig. 6.11.

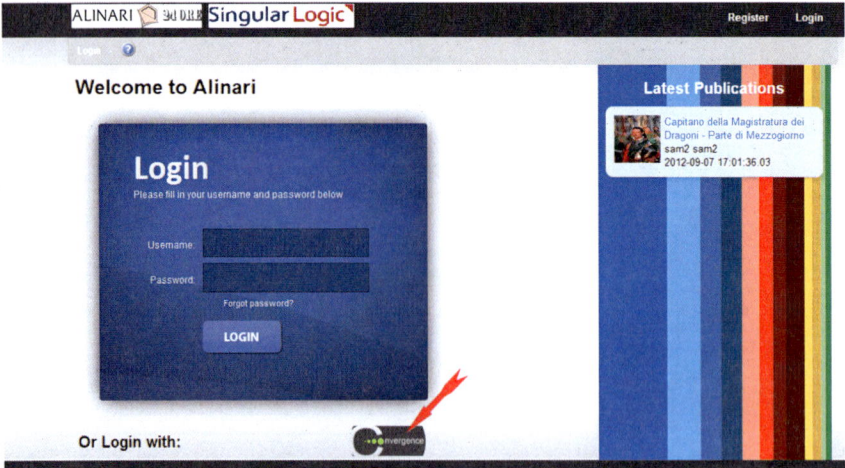

Fig. 6.9 Alinari application using SSO login

Fig. 6.10 OAuth server page for SSO card-based authentication

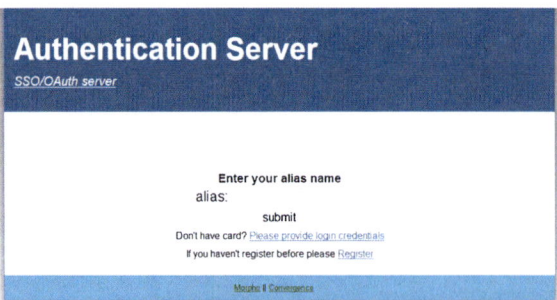

Fig. 6.11 SSO server page for SSO authentication based on user name and password

(iii)   There it is checked whether the user has already logged into the SSO server.

(iv)   If not logged in yet, the user will automatically be asked to authenticate.

(v)   If logged in the user is then redirected back to the SSO filter with user credential and signature, based on the secret key of the application.

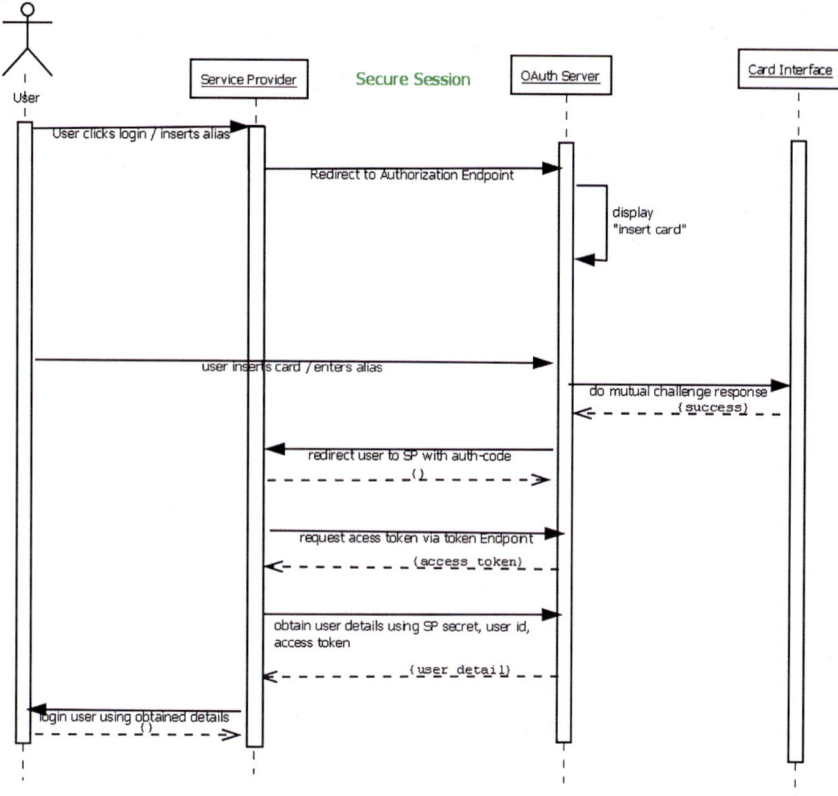

Fig. 6.12 Interaction diagram for OAuth authentication

(vi)   The SSO filter reads the user credentials to check if the user is logged in or not.

(vii) If the user is logged in, the session is set with user alias, first name and last name.

(viii) The SSO Filter redirects the control back to the application.

(ix)   Now the application knows from the session data set at (vii) whether the user is logged in or not.

6.6.2 SSO Authentication with OAuth

Our SSO server also supports the standard OAuth based authentication. Figure 6.12 shows the OAuth authentication (OAuth 2012). When the user clicks "Login with Convergence" in any application (cf. for instance the Alinari application in Fig. 6.9), he/she is redirected to a URL which is used as the common authorization

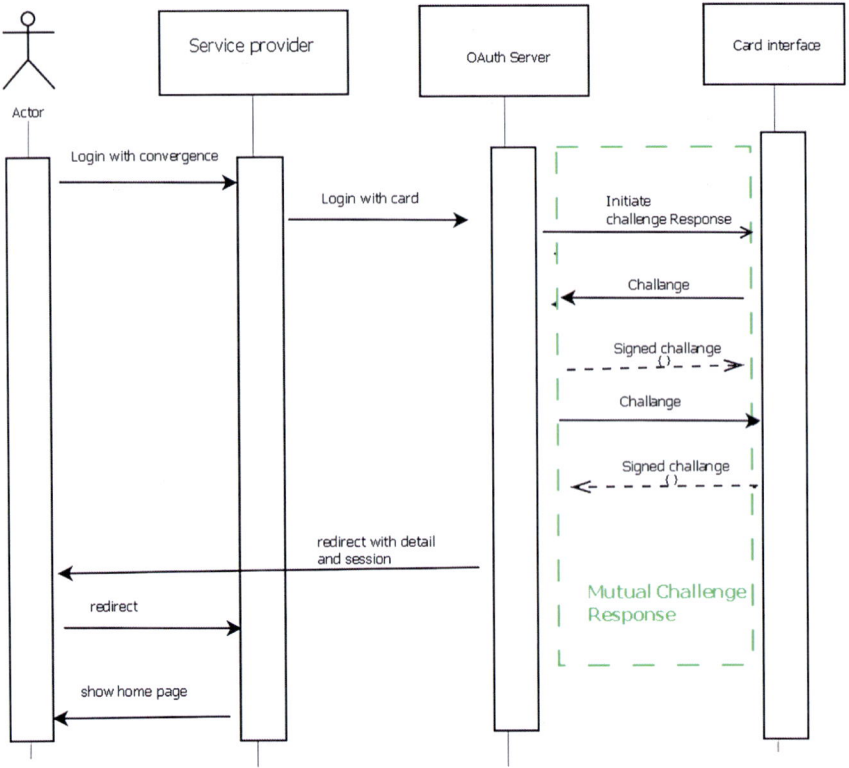

Fig. 6.13 Interaction diagram of mutual challenge-response

end point. There the user is prompted to enter his smart card. Then a mutual authentication between the card and the server takes place. After this the user is redirected back to the service provider with an OAuth code. This code along with the client id and client secret is used by the service provider to request for an access token. Once the access token is received, the service provider can use this access token to request further data, e.g. in our case, the service provider may request user detail with the access token. As a matter of fact, this access token has to be sent via a secure channel. Once user detail is received successfully the user gets logged in the service provider. Figure 6.13 depicts the challenge response protocol.

6.7 Conclusion

The Convergence system is based on a sound security framework providing all functions required in order to ensure data integrity, user and service authenticity as well as system integrity and security. We introduced an infrastructure with the core component CoSec. It has a distributed architecture with subcomponents realized on different computing platforms. CoSec provides standard methods of symmetric and asymmetric cryptography as well as advanced schemes such as anonymous group signatures and Identity-based encryption. A key feature of CoSec is smart card support for essential parts of all cryptographic primitives. We presented a couple of high level security functions for registration, authentication, signing, signature verification, encryption and decryption as well as proposals for secure Single-Sign-on and hybrid encryption. All these functions are embedded into the Convergence architecture. They are based on CoSec and provided to the Convergence middleware and network layer.

References

Allison B. Lewko, Brent Waters: Decentralizing Attribute-Based Encryption. EUROCRYPT 2011: 568–588.
Anadiotis A. C., Castro H., Charalampos P., Chiariglione L., Corlan L., Detti A., Gkonis P., Huebner T., Melazzi N. B., Mousas A., Ribas J., Salsano S., Sequeira D., Tanase M., Tropea G. (2011). Report: System Architecture. Convergence (D3.3).
Benoît Libert, Damien Vergnaud: Group Signatures with Verifier-Local Revo-cation and Backward Unlinkability in the Standard Model. CANS 2009: 498–517.
Carlisle Adams, Steve Lloyd, (1999): Understanding the public –key infrastructure. [cited 2013 May 15].
Crowd sourced. Phishing OAuth 2 [cited 2012 Nov 22]. Available online at: http://en.wikipedia.org/wiki/Phishing.
Dan Boneh, Xavier Boyen, Hovav Shacham: Short Group Signatures. CRYPTO 2004: 41–55.
David Galindo, Flavio D. Garcia: A Schnorr-Like Lightweight Identity-Based Signature Scheme. AFRICACRYPT 2009: 135–148.
Giorgio Calandriello, Panos Papadimitratos, Jean-Pierre Hubaux, Antonio Lioy: Efficient and Robust Pseudonymous Authentication in VANET. In Proceedings of the ACM International Workshop on Vehicular Ad Hoc Networks 2007.
John Bethencourt, Amit Sahai, Brent Waters: Ciphertext-Policy Attribute-Based Encryption. IEEE Symposium on Security and Privacy 2007: 321–334.
Julien Bringer, Hervé Chabanne, David Pointcheval, Sébastien Zimmer: An Application of the Boneh and Shacham Group Signature Scheme to Biometric Au-thentication. IWSEC 2008: 219–230.
Luther Martin: Introduction to Identity-Based Encryption. Artech House, 2008.
OAuth Consortium. OAuth Documentation. OAuth 2 [cited 2012 Nov 21]. Available online at: http://oauth.net/documentation/.

Chapter 7
The Adoption of Rights Expression Language in CONVERGENCE

Giuseppe Tropea, Giuseppe Bianchi, Nicola Blefari Melazzi,
Helder Castro, Leonardo Chiariglione, Angelo Difino,
Thomas Huebner, Angelos Christos-Anadiotis and Aziz Mousas

Abstract This chapter describes CONVERGENCE's licensing scheme and its
governance, based on the MPEG-21 part 5 standard and on the specific content
protection and rights management requirements, identified in the CONVERGENCE
use scenarios. In the digital media value chain, Rights Expression Languages
(RELs) are used to enable controlled access to digital resources, addressing several
different issues from the description of licenses to access and usage control,
payments, etc. A REL is an essential component of any security infrastructure
supporting differentiated controlled access to digital resources, and providing
adequate protection of intellectual property rights. Among these, the project has
selected the MPEG-21 part 5 open standard, which can be implemented in XML,
and is one of the main current contenders for a general-purpose REL. Our scheme is
designed in the light of CONVERGENCE's ability to distribute and manage any
kind of digital resource in a large distributed environment, and this chapter explains
how REL data is embedded into the CONVERGENCE data unit, the Versatile
Digital Item (VDI) and introduces a basic set of security features, based on digital
certificates, for the enforcement of the rights and conditions expressed in
CONVERGENCE licenses.

G. Tropea (✉) · G. Bianchi · N. Blefari Melazzi
Electronic Engineering Department, University of Rome "Tor Vergata", Rome, Italy
e-mail: giuseppe.tropea@cnit.it

H. Castro
INESC TEC, Faculty of Engineering, University of Porto, Porto, Portugal

L. Chiariglione · A. Difino
CEDEO.net, Turin, Italy

T. Huebner
MORPHO Cards, Paderborn, Germany

A. Christos-Anadiotis · A. Mousas
ICCS of National Technical University of Athens, Athens, Greece

F. Almeida et al. (eds.), *Enhancing the Internet with the CONVERGENCE System,*
Signals and Communication Technology, DOI: 10.1007/978-1-4471-5373-3_7,
© Springer-Verlag London 2014

7.1 Introduction

The explosive expansion of the Internet and spectacular associated developments in the computer industry have revolutionized the way people distribute and access content and information related services.

Until recent past, consumption of information was rigidly bound to physical objects containing the information (CDs, VHS cassettes, newspapers, etc.). The distribution infrastructure for these information-bearing objects (brick and mortar stores, newspaper shops, vending machines, etc.) was also physical, requiring heavy investment for start-up and maintenance. This investment came from dedicated economical entities (media retailers), which therefore became inescapable intermediaries between information creators and consumers.

The Internet and associated computer technologies have changed all this, enabling the "dematerialization" of information-carrying objects and the related distribution infrastructure. At the same time, the capabilities provided by household PCs have eliminated costs for the reproduction of information: with PCs, digital information objects can be reproduced without any consumption of physical resources. With the rise of the Internet, it has become possible to distribute information goods as purely digital objects, thereby drastically reducing distribution costs.

For the first time, information producers and consumers could *exchange information directly, without intermediaries* and at a very low cost.

These trends have fuelled the development of new technologies to facilitate, automate and manage content flow and service access over the Internet. The result has been an explosion in the exchange of legitimate (commercial and other legal ventures) and illegitimate (content piracy) media assets, which turned the Internet into the heart of the *digital economy*. However, harnessing its full potential requires new technologies for the management of *intellectual property rights*. If CONVERGENCE is to be a platform in the future Internet, this is an important issue. CONVERGENCE's technical structure and mode of operation has to safeguard the rights of all users, including final content consumers, original content producers and intermediating entities. In other words, CONVERGENCE has to manage and enforce access and use rights for the digital contents it distributes, independently of the terminal it connects (computers, mobile phones, other equipment) and independently whether the connection is through the Internet or via other telecommunication networks.

To satisfy these requirements within the CONVERGENCE environment a number of procedures must be observed: appropriate entities have to formally specify and authenticate relevant user rights; preferences must be formally specified and authenticated by appropriate entities; and the related information must be stored in information objects and services. These objects/services must then be securely stored/operated and made available to all authorized requiring entities. CONVERGENCE CoMid instances will use this information to enforce users' rights.

The formal specification of information pertaining to user rights and preferences, for the manipulation of digital items, is typically expressed in licenses. There already exist a number of tools for the specification of such licenses. The language in which a license is formulated is traditionally called a *Rights Expression Language (REL)*. Rights Expression Languages were pioneered by Mark Stefik at Xerox Labs, leading to the first (LISP-based) REL XrML (Stefik 1996). Modern REL standards include ODRL, CCREL, OMA DRM, and MPEG-21 REL.[1]

CONVERGENCE uses the MPEG-21 REL. This choice is motivated by the scope of the standard, the broad range of rights-related situations it can formally and precisely describe and the additional facilities offered by the rest of the MPEG-21 standard, notably the Digital Item Declaration concept (DID) the Rights Data Dictionary (RDD), the Intellectual Property Management and Protection specification (IPMP), etc. The great versatility and completeness of the set of MPEG-21 standards was thus instrumental in its selection as the core technology in CONVERGENCE. Additionally, it is relatively easy to define further functionality adopting the same principles, if deemed necessary to support specific CONVERGENCE requirements.

7.2 Structuring of VDIs in the Presence of REL Statements and Relationships

The MPEG-21 REL standard applies REL statements to resources contained within Digital Items. In the CONVERGENCE approach, on the other hand, data about resources (metadata, or descriptors of data) are as important as the resources themselves. To reconcile the differences between these approaches it is necessary to specify how CONVERGENCE will embed complex REL statements in VDIs.

Details of the structure of a VDI are reported in Chap. 5 of this book. Essentially, a VDI is conceived as follows:

```
.....
<item>
      <descriptor>
      <resource>
<item>
.....
```

This standard clearly represents the difference between data (sometimes called the *essence*) and data-about-data (usually called *metadata*), and it allows very complex groupings and nesting of Item(s). It is a flexible mechanism, which is

---

[1] For further information on these standards, see the links given in the references section of this chapter.

well-adapted to the human preference for two-level distinctions. However, it is also dangerous, since it leaves the door open to human interpretation of what to categorize as data and what as metadata.

When the concept of Digital Item was first introduced in MPEG, this danger was limited by the basic characteristics of the standard, in which:

- Descriptor is meant to hold simple information such as "The title of the song is 'Interstellar Overdrive'";
- Descriptor is not designed to be crawled by search engines;
- Relationships between Items were mostly specified by the grouping and nesting of physical Items.

As soon as the MPEG DI standard emerged, designers exploited the freedom provided by Descriptor tags, to enrich DI behavior. Likewise, CONVERGENCE, pushes the role of metadata Descriptors to their limits. In CONVERGENCE the situation is as follows:

- Descriptors hold valuable, structured information;
- Items are preferably shallow and not nested;
- Descriptors bear the load of linking distinct items into a structured web of relationships;
- such relationships are complex and semantically tagged;
- Descriptors are searchable;
- Descriptors are expressed in machine-readable schemas (ontologies).

The CONVERGENCE approach makes the distinction between data and metadata much more fuzzy. As an example, let us consider Samsam, a big hardware manufacturer, which is about to launch a new LED TV. The VDI that will be created to represent the TV needs to package the following pieces of information:

- name, brand, model;
- dimensions, weight;
- features;
- warranty;
- relationships with other similar products.

How should the different pieces of this information be packaged? Are they to be considered resources (essence data) or descriptors (metadata)?

The approach favored in CONVERGENCE is to package valuable information as generic resources and then to link resources by identifying which are data, and which contain information about that data. In the end, therefore, there is no encoded difference between resources and descriptors. All that exist are digital objects linked to other digital objects.

CONVERGENCE has further decided to make a distinction between the packages used for publication and search (Publication and Subscription VDIs, referred to as P-VDIs) and those used for self-standing resources (Resource VDIs, referred to as R-VDIs).

Using this scheme allows us to cope with our example in the following way:

- Warranty, hardware features, and the like, are packaged as Resources, in one or many R-VDIs;
- As these informations are resources, REL statements can be applied to them. They represent valuable information, which can be protected;
- When the resource is published in the semantic overlay (see Chaps. 2 and 4), making it discoverable, the relevant information is packaged as a Resource of the P-VDI (snippets of the features, model, relationships it is involved in);
- A REL license can be applied to the resource of the P-VDI making it possible to control who can read it (e.g., who can search for the features of the TV).

This is an example of an important CONVERGENCE requirement, namely the need to express and manage rights to what may be considered as metadata for (digital) resources. By representing (meta) data itself as resources/items, linked to the "original" object, it becomes possible to apply the full power of the REL. When (meta) data is intended to be searchable, it can be embedded as a resource inside a P-VDI.

This makes it possible to mark the syntactic distinction between Resource tags and Descriptor tags in more subtle ways than in the original standard. For instance:

- Descriptors can be treated as the "public part" of the package, holding simple descriptive sentences. The system retrieves a package, and automatically presents it to the user;
- Descriptors can be treated as the parts of the package to which REL is not applied. If metadata is valuable and people want to protect it, it can be packaged as additional Resources;
- Descriptors can be treated as the parts of the VDI that contain searchable hints and annotations, guiding the creation of publication VDIs out of resource VDIs;
- Descriptors can be used to carry system-level information such as relationships, expiry dates, sequence identifiers etc.

7.3 Control of Resources in VDIs

In CONVERGENCE, resources are referenced by VDIs. In many situations, the access to such resources needs to be controlled and hence this necessity naturally extends also to VDIs. The entities that can manipulate VDIs are either end users/devices, which consume resources, or peers of the system responsible for search and match operations. Different types of VDIs may circulate in the CONVERGENCE system, presenting different requirements for control and licensing. Nevertheless, it is possible to identify three major categories to group those different types of VDIs with specific control requirements. These requirements are therefore classified by type of VDI:

- *Resource VDIs*—R-VDIs represent/enclose actual consumable resources. They thus present the most pressing requirements for control and licensing. The requirements for R-VDIs are as follows:

  - The system has to control access to (consumption of) R-VDI resources, enforcing the corresponding licenses. Access should only be granted to authorized users.
  - The system has to control publication of R-VDI resources by users (through P-VDIs) enforcing the corresponding licenses. A resource may only be published by a specific P-VDI, if the user issuing the P-VDI has the right to publish the resource in question. This means that a different user can (re)publish the same R-VDI, with different metadata, only if she has the right to do so. The original creator can publish the same resource multiple times exposing different aspects of the resource. None of the rights expressed in the original R-VDI should prevent users from creating independent R-VDIs commenting, criticizing, or recommending a resource.
  - "Updating" of R-VDIs by new VDIs in the same sequence must be controlled by the system through the enforcement of the corresponding licenses. Only users authorized to manipulate the resource (typically the owner of the resource or the R-VDI creator) may update an R-VDI. This can be interpreted as the right to reuse the same sequence identifier for a newly created VDI.
  - Revocation of R-VDIs must be controlled by the system through the enforcement of the corresponding licenses. Only authorized users (typically the resource owner or R-VDI creator), may revoke an R-VDI.

- *Publication VDIs*—P-VDIs declare the existence of resources in the space of searchable objects by publishing their metadata and a link to the corresponding R-VDIs. The main control/licensing needs pertaining to this type of VDI are the following:

  - The system shall control access to (inspection of the contents of) P-VDI content (metadata) and matching of subscriptions to P-VDI metadata, enforcing the corresponding licenses. The only peers authorized to access its content are those belonging to the fractal (set of peers dealing with the same topic) where it was injected. Therefore, a search issued by a user will only find P-VDIs, which the user is entitled to access, and a peer will only perform a search operation for a P-VDI if it has the right to do so.
  - The system shall control revocation of P-VDIs through the enforcement of the corresponding governing licenses. Only authorized users may update a P-VDI (typically the P-VDI creator).
  - The system shall control reporting of subscription results to users, enforcing the corresponding licenses. Only authorized fractals may issue such reports and these may be directed only to an authorized set of users.

- *Subscription VDIs*—an S-VDI declares user's subscription criteria. The main control/licensing requirements for this type of VDIs are the following:

- The system shall control access to (inspection of the contents of) S-VDIs through the enforcement of the corresponding licenses. Only authorized fractals may process the contents of these licenses and only at the service of authorized users. This right can be used to regulate terms for collecting statistics about user subscriptions to specific VDIs.
- The system shall control revocation of S-VDIs through the enforcement of the corresponding licenses. Only authorized users (typically the S-VDI issuer), may revoke an S-VDI.

7.4 REL Verbs

The rights that an entity is entitled to when manipulating and interacting with resources, are expressed through a controlled language (the REL). Each standardized language specifies a limited set of verbs to indicate the type of actions that are allowed and the restrictions associated to those actions. Table 7.1 lists the twenty-one verbs and associated semantics that are defined in the MPEG-21 REL standard (ISO/IEC 21000-5: 2004) and its three amendments. This list provided the starting point when performing the analysis of the requirements in terms of usage rights of the CONVERGENCE use case scenarios.

7.5 Implementation with MPEG-21 REL

This section starts by presenting the general implementation principles of MPEG-21 REL licenses before proceeding to CONVERGENCE-specific implementation issues.

7.5.1 Overview

All the requirements outlined in previous section reflect a basic underlying need for *access control*. Controlling and restricting the ability to consume a media resource or to process a metadata resource, basically means limiting users' rights to access the resource in a specific way. In other words, the control requirements can be handled through the concession or denial of specific rights to users or fractals.

In real world situations, such as those described in the CONVERGENCE use scenarios, the "goods" that need to be delivered and controlled can be manipulated in different ways by different categories of user. For instance, resources of a media company such as Alinari may be consumed by a large number of authorized consumer users, and edited by a smaller set of editor users.

Table 7.1 Standardized ActType (verbs) supporting ISO/IEC 21000-5

ActType (Verb)	Definition
Adapt	To *ChangeTransiently* an existing *Resource* to *Derive* a new *Resource*
Delete	To *Destroy* a *DigitalResource*
Delist	The right to unlink or delist (the reference to) the related resource from a related playback control sequences description (i.e. play-list) for the optical disc when the play-list is newly created from an existing one
Diminish	To *Derive* a new *Resource* which is smaller than its *Source*
Embed	To put a *Resource* into another *Resource*
Enhance	To *Derive* a new *Resource* which is larger than its *Source*
Enlarge	To *Modify* a *Resource* by adding to it
Enlist	To link the related resource into a new playback control sequences description (i.e. play-list) for the optical disc
Execute	To execute a *DigitalResource*
Export	To export the associated broadcast program to another rendering or storage device
ExtendRights	To extend the rights which are the originally transmitted
GovernedAdapt	To adapt the resource and at the same time to result in certain rights being associated with the adapted resource
GovernedCopy	To copy the resource and at the same time to result in certain rights being associated to the copied resource
GovernedMove	To move the resource and at the same time to result in certain rights being associated to the moved resource
Install	To follow the instructions provided by an *InstallingResource*
Modify	To *Change* a *Resource*, preserving the alterations made
Move	To relocate a *Resource* from one *Place* to another
Play	To *Derive* a *Transient* and directly *Perceivable* representation of a *Resource*
Print	To *Derive* a *Fixed* and directly *Perceivable* representation of a *Resource*
Reduce	To *Modify* a *Resource* by taking away from it
Uninstall	To follow the instructions provided by an *UninstallingResource*

Issuing a specific license to every individual user for every good to which the user has some kind of right, would lead to a proliferation of licenses, and an increase in the burden of managing and enforcing them.

To avoid this, the CONVERGENCE licensing scheme attributes licenses and corresponding rights, not to individual users but to user groups. The only individual licenses will be those declaring that an individual belongs to a specific group. It follows that every scenario requires an entity (the Governance Entity), responsible for governing a specific class of content. This entity will then establish a set of "authority levels", each conferring a specific set of rights to a specific user group. For each level and individual Digital Item, an appropriately authenticated license will define the rights concerned and specify the user group to which they will be attributed.

Whenever a specific user acquires the right to belong to a specific user group, the Governance Entity will issue him/her a license, declaring that he/she is a member of the group. This will allow the CONVERGENCE system to enforce the licenses, assessing if a specific user is allowed to perform a specified action on a particular resource.

Figure 7.1 illustrates the rights attributed to four user groups, in which each group has a specific set of rights (or "rights area"), represented by a rectangle. The greater the area of the rights set, the broader the rights that the corresponding user group possesses. For instance, the group of users who have adhered only to the free service (Non-Paying UserGroup), has the least rights, while the group of the users which pay the highest fees (High-Paying UserGroup), and the group of editor users has the most rights.

In the situation described in Fig. 7.1, there is one license defining user group rights for each group, and for each resource to be controlled. Thus the Governance Entity would issue user John with an individual license declaring him to belong to user group "Low-Paying UserGroup" while editor Mary would receive an individual license declaring her to belong to user group "Editor UserGroup".

In this setting, fractals are also groups, this time composed of multiple peers. CONVERGENCE's licensing scheme can thus attribute rights to fractals, rather than to individual peers. Each individual peer would then be given a license, certifying that it belongs to a specific fractal. This situation is depicted in Fig. 7.2.

7.5.2 Base Mode

The governing and license scheme described above can be implemented in MPEG-21 REL through the two types of licenses presented in Fig. 7.3.

The right side of this image presents a User Group license. These licenses must:

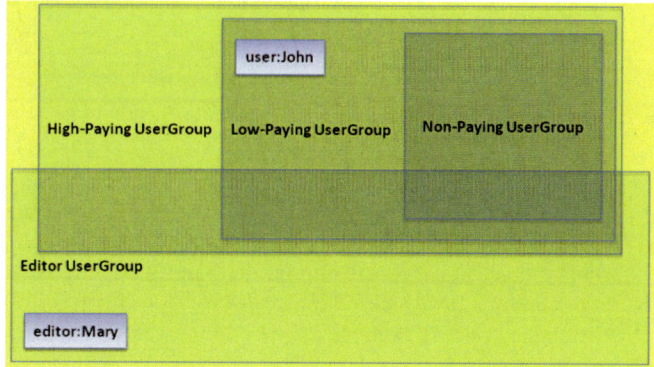

Fig. 7.1 Example of user groups and related "rights sets"

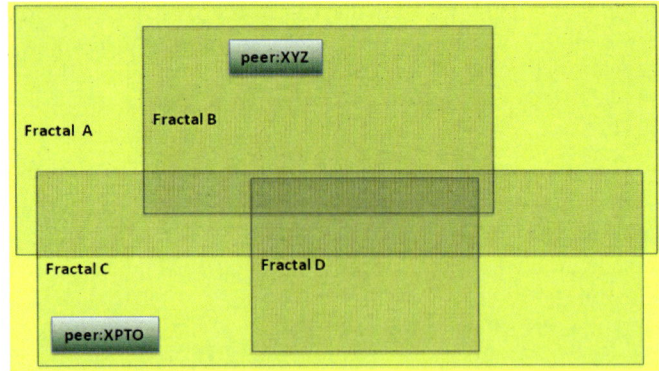

Fig. 7.2 Example of fractal "rights sets"

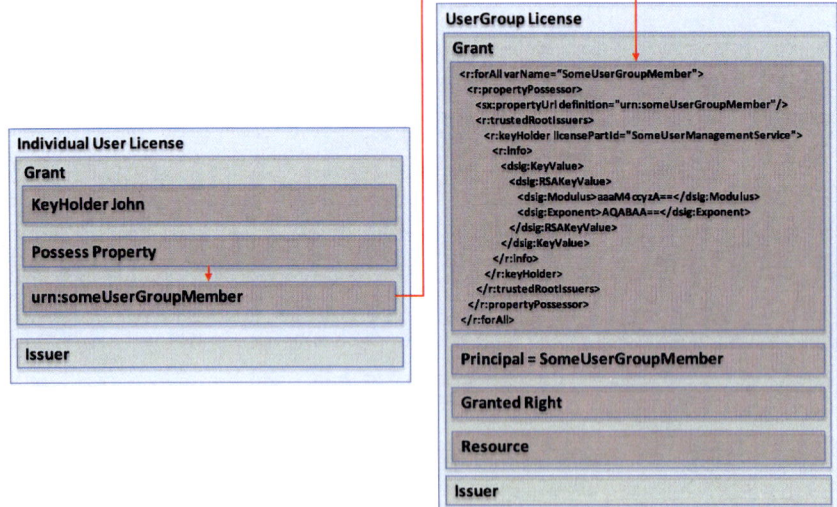

Fig. 7.3 CONVERGENCE main licenses

- contain a Grant which, in its turn:
  - defines a variable (through the forAll element), which basically stands for all entities which possess some specific property. In the given graphical example it stands for the entities possessing the property of "urn:someUserGroupMember", that is, the property of being a member of user group "someUserGroup";
  - defines the Principal (the entity to which rights are being granted) as any entity of the type defined in the previous variable;
  - defines the Right which is being granted by the license. This should be a right that is performable over a digital item;

- specifies the Resource over which the right is being granted (the Digital Item);
- identify the Issuer of the license. This should typically be an entity, which operates a specific "content channel", owns the content that it distributes and thus sets the content accessing policy (the specific Governance Entity).

The left side of that image presents the individual user's license (or User Membership License), which certifies his belonging to a specific user group. These licenses must:

- contain a Grant which, in its turn:
  - defines the Principal of the license. Employs the keyHolder element to uniquely identify a specific user.
  - defines the Right which is being granted by the license. This should be the possession of a specific property, which is defined (below) as the license's Resource.
  - specifies the Resource over which the right is being granted. This resource should be the property of belonging to some specific user group.
  - identify the Issuer of the license. This should typically be an entity which operates a specific "content channel", owns the content that it distributes and thus sets the content accessing policy, that is, specifies the different user groups and their corresponding right and privileges and manages users' participation in said user groups (the specific Governance Entity).

In the example given in Fig. 7.3, in the license on the left (the User Membership License), user (KeyHolder) John is granted a right which corresponds to the possession of the property "urn:someUserGroupMember", that is, the property of belonging to user group "someUserGroup".

In the license on the right (the UserGroup license), it is defined that all entities that possess the property "urn:someUserGroupMember" have a specific right over some specific resource.

Combining the two licenses, CONVERGENCE's content governance provisions are able to determine that John is entitled to exert the defined right over the specified resource.

In each CONVERGENCE scenario, the User Licenses (or User Membership Licenses) and UserGroup Licenses are issued by the appropriate Governance Entity. As they refer to one specific resource, UserGroup Licenses are generated at VDI creation time, and have to be embedded within the appropriate VDIs (as presented in Fig. 7.4). User Membership Licenses are generated whenever a specific user acquires the right to belong to a user group and are inserted in the user's VDI (as presented in Fig. 7.4).

The situation for the rights of fractals is very similar (please see Chap. 4 for the description of fractals within CONVERGENCE). Fractal rights are expressed in Fractal Licenses. The structure of the licenses and the production process are the same as for UserGroups Licenses. The only difference is that the Principals are fractals (and not groups of users), and that the granted rights are different.

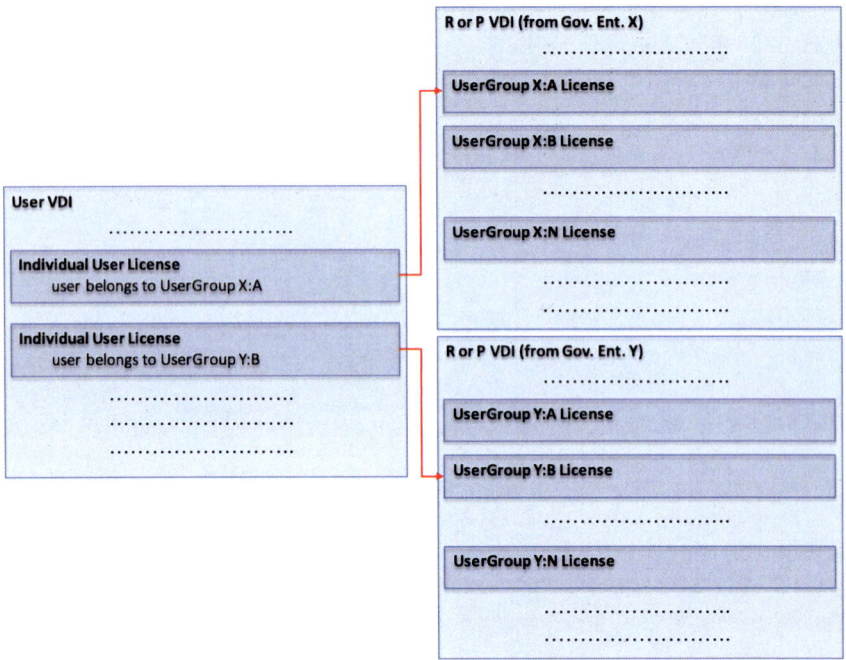

Fig. 7.4 License placement in VDIs

In this setting the equivalent of a User Membership License is a Peer Membership License. In this case, however, the structure of the license and the production process cannot be the same. Conceptually, individual Peers (Principals) are granted the right to belong to specific Fractals. However important distinctions break the symmetry. Fractals are highly dynamic in their nature, with Peers joining and leaving depending on the content they publish or subscribe. Moreover, it is impossible, given the size of the system, to maintain a centralized directory of the distribution of peers over fractals. Maintaining a decentralized, scalable system of "Fractals of Trust" is a major challenge. Such a system will need to guarantee the same kind of trust required for a Peer Membership License, but in a decentralized fashion. The following sections presents a preliminary design, which will require further work to achieve full maturity.

7.5.3 License Optimization

In the scheme defined previously, every VDI in the CONVERGENCE system needs to embed not only Fractal Licenses but also UserGroup licenses for all relevant user groups.

To reduce duplication, it would be possible to merge all the grants, contained in different licenses, into a single license, as shown on the left side of Fig. 7.5. The same procedure could also be applied to Fractal Licenses.

The license would thus contain a grant for each relevant user group (or fractal) and each grant would use its resource declaring element to reference the resource whose manipulation is controlled by the license. The resource would be declared only once, in the Inventory element, at the beginning of the license.

However, even in this scheme, the license would contain potentially a huge number of different grants (number of rights $*\times$ number of UserGroups). There is thus room for further optimization, providing an optimally ordered and minimally redundant internal structure for the license.

Each user group may hold more than one right over a particular digital resource. This means that Grants belonging to the same Principal (which may only specify one Right each) can be packed together, using the grant Group element.

Thus, let us assume the UserGroups license contains:

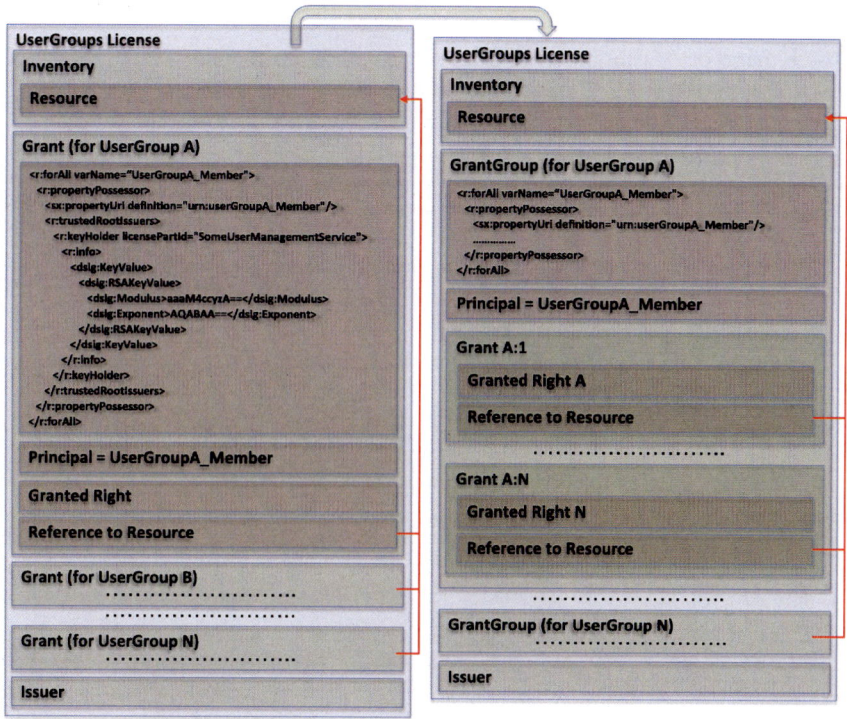

Fig. 7.5 Usergroup licenses merged into a single usergroups license

- one GrantGroup, for each UserGroup, and the GrantGroup contains:

  - one declaration of the Principal defined in the form of a variable which represents all entities that possess a specific property (the property of belonging to a specific user group);
  - one Grant for each of the principal's rights over the controlled resource.

Then the resulting license (presented on the right side of Fig. 7.5) will define all the rights of all the user groups over the controlled resource. Equivalently, a Fractals License can define the rights of all relevant Fractals over the controlled resource.

7.5.4 Expression of Some Specific Rights

In the light of the requirements described in Sect. 7.3, it will be necessary to control at least the following actions:

- Regarding the R-VDI:

  - Playing or consuming the media resources of the R-VDI.
  - Deleting the R-VDI's resources (and consequently the R-VDI itself) from the CONVERGENCE system.
  - Revoking an old R-VDI and storing a new R-VDI.
- Regarding the P-VDI:

  - Performing matches on the metadata resources of the P-VDI.
  - Issuing event reports regarding the previous matches.
  - Collecting information from the metadata contained in the S-VDI, for statistical purposes.
  - Revoking the P-VDI from the CONVERGENCE system.
  - Revoking an old P-VDI and storing a new P-VDI.
- Regarding the S-VDI:

  - Collecting information from the query contained in the S-VDI, to aggregate it for statistical purposes.
  - Revoking the S-VDI from the CONVERGENCE system.

The focus of the following discussions will be on licenses for publications and subscriptions.

Not all of the actions just described have a corresponding expression in the MPEG-21 REL. Table 7.2 below, provides a summary description of actions not considered in the standard.

Table 7.2 Corresponding MPEG-21 actions

Action	Definition
Match	The right (of Fractals), to store a set of metadata (P-VDI case) or a query formulation (S-VDI case), as the payload of a *did:Resource* element of the VDI, and to perform a matching of said resource against complementary resources, belonging to a specified set of users
Notify	The right (of Fractals), to report, (to a specific set of users), a match between a specific *Resource*, (specific metadata contents of a *did:Resource* of a VDI), and a user query or user subscription
Aggregate	The right of Peers (or crawlers external to the CONVERGENCE middleware), to report aggregated information about Rights that are executed on specific *Resources* (i.e. specific metadata contents of a *did:Resource* of a P-VDI, a specific user query or user subscription formulation of a S-VDI) to a specific set of users

Thus, generally speaking, licenses in publications and subscriptions, control:

• what is to be collected or matched to what;
• in which way such information is to be distributed to whom;
• until when.

All three actions listed in the table above require the specification of a list of users. Our preliminary solution is to include this list in the Condition fields of the license. However, this proposal needs to be validated by further investigations.

At first, it might seem that Match and Notify are so tightly connected that only one of them is needed. However, in reality it is necessary to treat and express both operations as separate rights. The reason stems from the situation depicted in Fig. 7.6.

The scenario illustrated in the figure, refers to the case where a single user creates a subscription VDI indicating a certain matching criteria and a group of users to whom notifications should be sent. Thus, the aim is for the system to send notifications to a group of users, who did not directly subscribe to the indicated criteria, when matching P-VDIs are found by the system. This situation could occur, for example, with a retailer who wanted their customers to be automatically notified of promotions in his/her store. Let us assume that this retailer is UserA. Figure 7.4 shows an S-VDI created by UserA, embedding MPEG-21 ERRs (Event Report Requests) to specify the matching criteria for the subscription, as well as the set of target users to be notified with an ER (Event Report), when matching P-VDIs are found (ReqNotifUserSet).

ReqMatchUserSet is the set of users who have indicated a specific match, which in this case, contains only one single user (i.e., UserA). On the other hand, ReqNotifUserSet contains other users besides UserA, which means that ReqNotifUserSet ≠ ReqMatchUserSet. This is the situation where the subscribing user is performing not only a traditional subscription for him/herself, but also on behalf of other users. Basically he/she is performing a push of notifications to third party users.

Another possible situation that illustrates the difference between "match" and "notify", in terms of usage rights, is the case where a content publishing user may

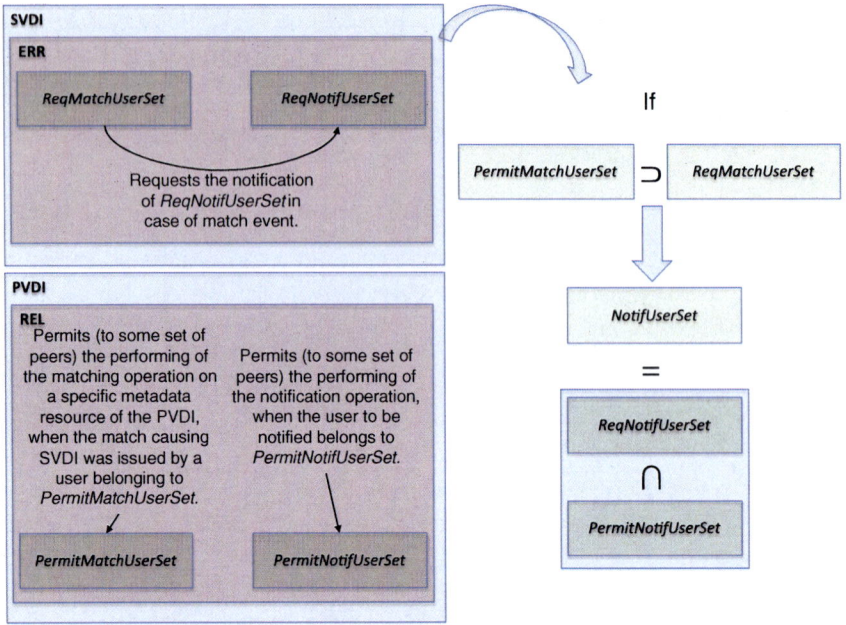

Fig. 7.6 Event reporting and P-VDI rights scenario

wish to limit the set of users whose S-VDIs are matched against his/her P-VDIs (PermitMatchUserSet), or to limit the set of users who receive match notifications for his/her P-VDIs (PermitNotifUserSet).

If the condition ReqNotifUserSet = ReqMatchUserSet was mandatory for all situations, the only user who could receive the ER would be the original subscriber. Thus the Match and Notify operations would involve the same sets of users and, in terms of rights management, they could be treated in a joint manner.

However, CONVERGENCE also supports the condition when ReqNotifUserSet $\neq$ ReqMatchUserSet. Given that in this case the Match and Notify operations concern different sets of users, they must be treated separately in terms of rights management. In other words, PermitMatchUserSet $\neq$ PermitNotifUserSet.

Thus, the publishing user should be able to set Match and Notify rights separately and associate them with different sets of user (PermitMatchUserSet, PermitNotifUserSet).

The set of users who will actually receive an ER from a specific Match event (NotifUserSet), is determined as follows:

- if PermitMatchUserSet $\supset$ ReqMatchUserSet:

  – the intersection—ReqNotifUserSet $\cap$ PermitNotifUserSet.
- else:

  – the empty set—$\varnothing$.

To avoid pollution of the system by a myriad of notifications for every match, the system will assume PermitNotifUserSet = PermitMatchUserSet, unless a publisher specifies a different PermitNotifUserSet. In other words all users whose subscriptions match the P-VDI may receive a notification.

Similarly, the system will assume that PermitMatchUserSet = ReqMatchUserSet unless a publisher specifies a different PermitMatchUserSet. Thus any user's subscription can be matched against the P-VDI, but only the subscribing user has the right to be notified of the match.

Match and Notify rights can be expressed in UserGroups licenses, (or Fractal licenses) using the rightUri REL element, as illustrated below.

```
.....
<sx:rightUri definition="conv:right:match"/>
<sx:rightUri definition="conv:right:notify"/>
<sx:rightUri definition="conv:right:aggregate"/>
.....
```

An alternative would be to develop a CONVERGENCE-specific XML schema in which these rights would be expressed by extending the REL Right type, as illustrated below.

```
.....
<xsd:complexType name="Match">
  <xsd:complexContent>
    <xsd:extension base="r:Right"/>
  </xsd:complexContent>
</xsd:complexType>
.....
```

7.6 Techniques to Implement REL in CONVERGENCE

7.6.1 Controlling the Matching Process

An issue that has been debated at length during the design of inter-VDI relationships is the governance of these relationships. Some of the devised use-cases have raised concerns such as those illustrated by the example here below.

A manufacturer releases a VDI for a product to a supplier. The supplier signs a deal with a retailer to sell the product. The retailer will now create a VDI, advertising his sales conditions, linking back to the original product VDI through the kind of semantic relationship described in Chap. 5 and fully specified in (Deliverable D4.1 2011). How can the original manufacturer control such a relationship, when there is no direct business connection between the manufacturer and the retailer? In other words, how can the system forbid a black market reseller from linking to the manufacturer's resource VDI claiming to be the official reseller or setting the recommended price for the product in a given city?

At first sight it seems that it would be useful to provide the manufacturer with this sort of control. Such a mechanism, if technically feasible, would provide:

- better control over the flow of information through the network;
- a new incentive for users to adopt the platform;
- an obstacle to false claims.

Further analysis suggests that users should be free to link their opinions (comments, annotations) to a VDI but that it might be desirable to prevent them from claiming an official/business relationship with the VDI owner. In the case just described, a manufacturer would not be interested in a technology that allowed a third party to make false claims about his product—and provided no way to check the authenticity of the information disclosed. At the same time, however, strong controls could pose a danger for freedom of expression.

CONVERGENCE therefore proposes a solution that does not encourage censorship but which maintains the advantages of control. This solution consists in giving to users the option of formulating subscription requests that filter out resources that do not come from certified sources.

In other words, CONVERGENCE offers a ranking and filtering complement to the Match TE, making it possible to distinguish genuine items from items that people might have published in bad faith, matching only VDIs that are signed by a specific person/entity/company.

This solution would limit misbehaviour and limit some denial of service attacks, removing the incentive for dishonest users to flood the network with fake VDIs.

This kind of mechanism could be used to ensure that children will always download genuine cartoons and do not run the risk of reaching inappropriate content, which could be built into software as in parental control systems. This would not remove users' technical freedom to attach improper (or undesirable, or, conversely, critical) content to the resources of reputable brands, but it would make it possible to filter out this content when needed.

The following section describes technical aspects of this approach, and the specific issues it raises.

7.6.1.1 Searching for Certified Content

Consolidated tools such as Public Key Certificates (PKC), Attribute Certificates (AC) and digital signatures already allow to filter out uncertified content. For example, AC technology (IETF RFC 3281, 2002) allows a company to certify that a person/business is an authorized reseller for an area, and that it has the right to set the price of its products.

In the following example Samsam, the producer of the novel BrightStar TV, has certified Novelties Inc. as a Certified_Partner for all its products. In this setting "Certified_Partner" is an attribute associated to Novelties Inc. in an AC issued by Samsam.

Samsam creates and publishes a Product VDI for the TV, packaging the specifications and maybe a detailed brochure.

Novelties Inc. has subscribed to all Samsam products. Once it receives the Product VDI for the new line of TV sets, it creates a Reseller VDI that packages special offers, a map to find the Novelties Inc. shop and a link to the website. A semantic relationship links the Reseller VDI to the Product VDI. Once this is done, Novelties Inc. creates and publishes a P-VDI letting users find the Reseller VDI, and advertising a recommended retail price. In other words, the P-VDI basically sets a price for the TV set and publishes it in the semantic overlay of CONVERGENCE.

The content of the P-VDI is expressed as a set of RDF statements (see Chap. 2), as shown below:

```
{

    NOVELTIES_R-VDI isResellerOf BRIGHTSTAR_R-VDI,
    BRIGHTSTAR_R-VDI hasPrice 300,
}
```

In our solution, Novelties associates the first RDF statement with the AC they have received from Samsam. Thus the P-VDI is packaged as follows:

- RDF statements block.
- Attributes Certificate for the Certified_Partner attribute.
- Public Key Certificate (for instance issued by VeriSign) for Novelties' digital identity.

This package, signed by Novelties. is sent to the peers of the semantic overlay.

User Nicola creates a subscription that asks: "Things that are isResellerOf BrightStar". CONVERGENCE matches all R-VDIs declaring the owner to be a resellers of the product.

But if Nicola formulates the query as follows: "Things that are isResellerOf BrightStar AND property isResellerOf certified by Samsam", and packages the S-VDI as follows:

- SPARQL equivalent of the above query.
- PKC of Samsam trusted by Nicola (for instance issued by VeriSign).

This allows CONVERGENCE to filter out bogus resellers and match Novelties Inc. only.

In practice, Novelties has backed up its statement that it is a reseller (the isResellerOf property) with a certificate, from Samsam, for a somewhat different and specific attribute (i.e. "Certified_Partner"). Thus Samsam has not directly certified either the isResellerOf property, or the "NOVELTIES_R-VDI isResellerOf BRIGHTSTAR_R-VDI" statement. In other words Samsam is endorsing Novelties as a Certified_Partner, not as a Reseller (nor generic nor specific for reselling BrighStar products). It is Novelties' initiative to endorse that statement with the Certified_Partner AC. In other words, it is not Samsam who is establishing a synonymy between Certified_Partner and isResellerOf, but Novelties. It is the end-user who decides whether or not to accept the association.

Nicola will receive notifications that contain an isResellerOf property, which has been coupled, by the creator of the P-VDI, with a legitimate AC certificate issued by Samsam. He will thus know that:

- The P-VDI has been created by Novelties Inc;
- That it has not been tampered with;
- That Novelties Inc. possess "Certified_Partner" certification from Samsam;
- That it is it indeed Samsam Corp. that has issued the certificate;
- That Novelties Inc. wishes to endorse its status as a reseller of BrightStar tv sets with said certificate.

It is up to Nicola whether to accept the notification and to use the associated Reseller VDI (that then links back to the Product VDI). For this reason, the system flags the answer with a warning that the certificate is for attribute "Certified_Partner" while the declared associated property is "isResellerOf". A properly designed Application should allow Nicola to check with the certified attributes with his own eyes, and check them against the claims that are being made.

Of course a simplified approach is possible, as explained later, where the user just asks for content certified by Samsam, without specifying a precise attribute to look for in the query. In this case the user does not need to confirm the matching of strings, because he has declared Samsam as trusted by him once and for all.

Reasoning along the same lines, Novelties Inc. could associate the same Certified_Partner, to the "BRIGHTSTAR_R-VDI hasPrice 300" statement. This would be perfectly legitimate.

In this scenario, if Nicola requests "BrightStar that hasPrice 300 AND property hasPrice certified by Samsam", he will get the Novelties Inc. resource again. In this case, however, the system shows that his request for a certified hasPrice is backed up by Novelties only by means of a vague "Certified_Partner" attribute from Samsam, he might decide to discard Novelties as non-trustworthy. Or he might accept the Novelties claim. It is worth remembering that a match can occur in several peers of the same overlay fractal. There is thus a risk that the same subscription could yield multiple potentially useless results.

Moving a step further in the design process, it is possible to see that it might be the interest of Samsam to create a service that can check what properties/relationships Samsam considers as equivalent to the Certified_Partner attribute. Nicola or an application working on his behalf could then contact a service and find out that, for Samsam, Certified_Partner only implies hasRetailPrice.

In other words Samsam could deploy a private service based on CDS (the system's Common Dictionary Service, see other chapters) that expands a generic hasRetailPrice relationship into a "Certified_Partner" one and Nicola could add the service to the list of default CDS providers he uses in his private matching process. His private CDS can then use Samsam's standards to expand his queries.

However, a public distributed matching system such as CONVERGENCE's semantic overlay has two goals:

- To avoid a private company like Samsam from taking control of the "public" CDS by dictating which one of its certified relationships match English words or ontology properties that users might use in generic queries.
- To scale the matching process to a large number of peers, avoiding the need for a user to contact an external third party CDS to validate attributes contained in certificates, when comparing publications and subscriptions.

In theory, the distributed nature of CONVERGENCE's middleware would make it possible to deploy a broad range of solutions. On the one hand filtering could be performed locally by the end-user device, which would have to handle a flood of answers from the network and filter them afterwards. On the other hand core peers could perform the filtering before results are notified to the end-user, using public and common dictionaries. Many intermediate solutions are also possible.

Filtering by peers in the overlay would obviously be desirable. In this way, however, the network would not be able to exploit the user's own trusted synonyms.

A conceptually straightforward solution to the problem would be to:

- Clearly distinguish public, common CDS repositories with respect to private CDS repositories;
- Only use common CDS when expanding terms during distributed match;
- Allow for a subset of private expansion rules be packaged together with the S-VDI;
- Use those suggestions from the user, extracted from her private CDS repository, to complete the matching process.

Let us continue with the BrightStar example.

Alina has installed a private CDS repository on her device, sourcing expansion rules from Samsam's knowledge base. Now Alina formulates the same query Nicola formulated previously: "BrightStar that hasPrice 300 AND property hasPrice certified by Samsam". Her device packages the S-VDI as follows:

- SPARQL equivalent of the above query;
- Samsam PKC trusted by Alina (for instance issued by VeriSign);
- Snippet of expansion rule that declares: "Certified_Partner" only implies "hasRetailPrice".

Peers that receive this subscription can now filter out publications from Novelties Inc. declaring that hasPrice is backed up by Certified_Partner certification.

Another very important case, covered by this design, is the case when the user has no knowledge of the specific ACs Samsam has issued, and formulates a broad query such as: "Content that is endorsed by Samsam". This subscription does not match P-VDIs that are signed by Samsam, but P-VDIs signed by individuals who are holders of ACs issued by Samsam, regardless of the attribute string.

In summary, the solution described in this section guarantees that semantic expansion of query terms, a crucial part of CONVERGENCE match technology, is secure against infection by private information which could distort the results provided to the general public, but still allows private expansions.

The crucial issue here is governance of CDS repositories. In other words, who decides when and under what conditions a CDS repository is authoritative enough to be considered of public and common interest.

As a concluding remark, it is possible to observe that the solution proposed in CONVERGENCE confirms the decision that VDIs should be self-contained in terms of assertion of trust. S-VDIs and P-VDIs carry with them all certificates necessary for verification algorithms to run offline and in parallel in the core of the semantic network of CONVERGENCE. Distributed networks often have trade-offs between CPU time and the bandwidth required by the network overlay. The self-contained chain of trust within VDIs makes it possible to explore the terms of this trade-off, dropping any dependency on external services when certifying content.

This solution does not prevent anybody from linking to "official resources unless some conditions are met", but lets subscribers verify that content is endorsed by sources they trust.

7.6.2 A Secure Environment for CONVERGENCE Technologies

In previous chapters the power and flexibility of the REL was demonstrated. Nevertheless, it is obvious that the critical issue is the actual enforcement of the rights, described by REL. The only way to achieve this is to run the whole Security TE, and possibly other middleware technologies, within a trusted sandbox on the devices acting as CONVERGENCE peers.

To better understand the issues involved, it is sufficient to imagine the situation where an end-user in an unprotected terminal, which downloads a photo VDI containing a license to Play (view) an encrypted photo. The CONVERGENCE peer, based on the Play rights expressed in the license, decrypts and presents the photo to the user. However, given that the middleware runs in an unprotected environment, the peer could also be tricked to perform other type of manipulations on the photo, such as Store, Adapt etc., even though these rights have not been granted in the original license.

Hence, a complete enforcement of licenses can only be achieved when assuming a trusted "sandbox" device. To enforce a license of a content provider, he/she has to trust the sandbox to honor that license. The owner of the device, who wishes to "consume" the content, might in reality be glad if restrictive licenses were not honored.

CONVERGENCE deployments that can count on peers equipped with fully-fledged "Trusted Computing" systems using a Trusted Platform Module (TPM)-like secure hardware, will be able to check that the middleware behaves validly. Since management and presentation of VDIs is delegated to media engines at the middleware level and mediated via the Security TE, it is possible to limit the user's actions on resources he/she has no right to manipulate further.

On the other hand, enforcement can also be achieved in an open environment, through the use of a user's smartcard, which constitutes a small and limited, yet very trustworthy, "sandbox". In such a deployment, the smart card performs the crucial step in content decryption, namely the "unwrapping" of the content-key.

The security aspects of CONVERGENCE are in Chap. 6. In that chapter it is discussed support of REL in open environments using smart cards. The present chapter describes a promising approach explored by CONVERGENCE, based on so-called Attribute Based Encryption (ABE). Further technical details on how ABE is integrated into CONVERGENCE, are also been provided in Chap. 6, in particular Sect. 6.4.

7.6.3 State of the Art of the ABE Technology

Attribute Based Encryption (ABE) is an elegant form of key escrow allowing an issuer of content (or in general, the sender of a message) to encode specific access rules into his "ciphertext" (i.e., the encrypted content). ABE allows for an attribute-based access structure with logical gates like AND- and OR-gates (even threshold-gates), so that an issuer can encode complex decryption policies using these gates.

ABE can thus be described as a type of encrypted access control, where access control policies are either embedded in the user private keys or in the ciphertexts. An example taken from (Bethencourt et al. 2007) will help clarify the approach.

A head FBI agent may want to encrypt a sensitive memo so that only personnel who have certain credentials or attributes can access it. For instance, the head agent may specify the following structure for accessing this information:

"Public Corruption Office" AND ("Knoxville" OR "San Francisco") OR (management-level >5) OR ("Name: Charlie Eppes").

With this statement, the head agent had the intention to indicate that only the following entities should be allowed to see the memo: (1) agents who work at the public corruption offices at Knoxville or San Francisco; (2) FBI officials with a very high rank in the management chain; and (3) a consultant named Charlie Eppes.

In this example, ABE techniques can guarantee that agents not satisfying the encoded attributes are not able to decrypt the memo.

The general "high-level" setup is quite easy to describe: To begin with, (like with IBE, Identity Based Encryption) there is the need for an absolutely trustworthy master instance (a CA, "Central Authority") which generates system parameters, holds one master secret key it never reveals to anyone, and issues one common public key for the ABE-instance at hand. (Note the difference with IBE, where each user is associated with a public key, or even several public keys, of his own). For each member of the ABE-instance, the CA generates an individual private key consisting of components with a list of attributes "woven" into them.

A "message" is encrypted by encoding a decryption policy described through an "access structure" built up from logical combinations of attributes. A user can decrypt the corresponding "ciphertext" if and only if the access tree used for its encryption matches the sets of attributes woven into his or her private key.

ABE also tackles a fundamental problem immediately rising from such techniques: namely collusion. Two or more users, whose individual private keys may not satisfy the access structure of a given ciphertext, may still collude ("combine their private key material") if the union of attributes allocated to them allows decryption.

A simple example: user1 is >18 years old, but not a member of clubxy; user2 is a member, but <18 years. Take a message encrypted under an access structure requiring any receiver to be member in the clubxy and >18 years old. Neither user1 nor user2 satisfy both attributes, but joining together they do.

In reality, ABE is collusion resistant: the attributes "woven" into the private key components are "bound" to each other by encoding them with user-specific random numbers. Thereby attribute related components from different keys are incompatible. During decryption the embedded random numbers will only "cancel out" if compatible key components (with identical embedded random numbers) are used.

In plain words, a private key can only decrypt a ciphertext if its underlying access structure is satisfied by the attributes associated to components within one and the same private key. As a consequence, disjoint colluding users cannot simply "merge" their private keys to obtain a "combined" private key capable of decrypting a ciphertext whose access structure would only be satisfied by their joint set of attributes, while none of the colluding users could alone decrypt it. This means that in a particular ABE-instance, there is no need for all recipients to get their individual private keys from the CA.

7.6.3.1 Existing ABE Schemes

The CP-ABE scheme by Bethencourt et al. (2007) cited above, describes a scheme requiring one trusted CA for each ABE instance holding a master secret.

It is this CA that is responsible for governance and in particular for the derivation of decryption keys. For each such decryption key generation, the CA takes a set of attributes and outputs a key that identifies that set. Given that the master key is required, only the CA can generate such keys.

To avoid offering a single point of attack, the CA can be split into different CAs using secret sharing techniques. Some schemes even lend themselves to homomorphic encryption, making it possible to derive individual keys using secret shares and yet prevent recovery of the master secret.

Another very interesting ABE-variant uses a "decentralized" approach. "Decentralizing Attribute Based Encryption" (Lewko and Waters 2011), proposes a scheme that, unlike the schemes discussed so far, has no single master secret.

This decentralized approach allows much more flexibility: various trusted authorities can share attribute administration, with each authority retaining responsibility for "its" attributes. Global parameters exist, but they are all public, and there is no need for one master secret, or one single authority to be absolutely trustworthy. Users can encrypt designated messages to attribute-based access structures each involving various attributes from various authorities. Including users' trustworthy global IDs in key components makes the scheme collusion resistant.

Simply speaking, each user "collects" his/her key components from various authorities according to the attributes assigned to him/her. These key components are "masked" with the user's global ID. The mask cancels out during decryption of a message only if the encoded access structure is satisfied by attributes for which the receiving user possesses all necessary components (i.e. all masked with "his/her" ID, though possibly arising from different authorities). By contrast, colluding users can only combine components masked with different IDs, which will not cancel out.

7.6.4 Challenges in ABE Support to REL

In ABE, attributes are usually thought of as describing properties of recipients rather than attributes of the content itself. However the strong point of ABE is that basically any (binary) string can become an attribute. This is achieved by invoking a suitable hash-function mapping arbitrary strings to elements of a specific group.

The construction of the group involves mathematics; usually it is a prime order subgroup of an elliptic curve. Access trees have attributes assigned to their leaf-nodes, while each interior node works as a threshold gate determining the minimum number of its child nodes, which need to be satisfied.

Although, generically speaking, keys can be generated according to designated sets of attributes, further research would be needed to have ABE-based techniques fully integrated with the current architecture.

7.6.4.1 Mapping Complex REL Statements to ABE

Mapping complex REL statements to ABE requires a specific method for transferring REL descriptions to ABE. A promising approach, and one explored by CONVERGENCE, is to allow users to access portions of VDIs only if they possess a certain set of credentials or attributes.

Encrypting specific parts of content—portions of VDIs—would provide the ability to administer complex rights. For instance, it could be possible to allow free access to low-resolution versions of photos, whilst imposing additional decryption to decode the high-resolution components.

As an example, the next Table (Table 7.3) uses the specifically CONVERGENCE-developed usage scenario FMSH, to illustrate how ABE can be used to code the right to download a video (please see Chap. 8 with details on the CONVERGENCE applications).

Success in this task demands a clean and simple interface that lets us deal with two fundamental aspects of ABE techniques in a technology-agnostic way:

• Encrypt/decrypt operations;
• Procedures for releasing attributes and for interaction with one (or more) authorities. The interface should provide functionalities for temporal management of attributes, for instance to check if attributes are still valid and to issue replacements. Ideally it should also deal with attributes' revocation, but this is technically very much challenging since most ABE solutions do not support such a concept.

The interface would encapsulate the specific details of the ABE solution chosen, hiding them from CONVERGENCE technologies.

Table 7.3 Use of ABE in FMSH scenario

User	Conditions on user characteristics
FMSH analysts	(FMSH certified) AND (university level > = M1)
FMSH broadcaster	(FMSH certified) AND (localized in Paris, France)
FMSH VCO	(FMSH certified) AND (paid a 1 € fee)
INC	(INC certified)
Peruvian government	(Peruvian government certified) OR (localized in Peru)
Summary	((FMSH certified) AND ((university level $\geq$M1) OR (localized in Paris, France) OR (paid a 1 € fee))) OR (INC certified) OR (Peruvian government certified) OR (localized in Peru)

7.6.4.2 Coping with a Highly Decentralized System

As it was described in the previous sections, there are two approaches for distributed ABE solutions: shared-secret among multiple authorities, and truly decentralized techniques.

The problem with shared-secret approaches relates primarily to the management of the system, rather than to any intrinsic limitation of the shared-secret concept. There are distributed key generation algorithms (Pedersen 1991) that make it possible to easily build a shared-secret that is not known by the participants, and that allows verification in a way that participants cannot fake. The main problem lies in the steps authorities have to take to make the system work. This problem can only be alleviated by dynamic incremental setup techniques.

The decentralized approach from Lewko and Waters (2011) does not involve any shared-secret, hence avoiding setup procedures, constituting a major asset of this approach. It should be however noted, that any setup procedures required by a specific ABE scheme will be hidden from CONVERGENCE.

Furthermore, Lewko's decentralized technique suffers from three criticalities:

- It is extremely difficult to integrate revocation methods. To the best of our knowledge there are no known solutions for this. This is a problem common to all decentralized techniques. The only schemes that offer limited opportunities for revocation are single authority schemes.
- There is no formal proof that Prime groups are secure. This can be proved for composite groups. However their huge size makes them impractical. Of course, this does not mean the scheme is insecure, only that a formal proof is missing.
- Performance and overhead is problematic. This technique requires the addition of a point to the elliptic group, plus three more points per attribute. Thus, with 10 attributes in a policy, this sums up to 31. The technique also requires a high number of elliptic exponentiations for each attribute. Investigations on this issue are important for the advancement of the state-of-the-art.

These three points, while not extremely critical, are a challenge for the distributed technique. This implies that, if well-thought and implemented shared-secret schemes compensate such criticalities, it could be worth paying the price of the required setup phases.

Another example shows the difficulty of choosing the right ABE scheme for CONVERGENCE. To the best of our knowledge there are currently no short ciphertext techniques (i.e. schemes in which size is independent of the number of attributes) for multiple authority schemes. To date this has been achieved only for a single authority. It is likely that, if and when such techniques are developed, extensions will apply to multiple authorities' schemes and not to decentralized ones.

7.6.4.3 Dealing with Post-decryption Rights

How should rights, such as "read only" / "read + modify" / "store", etc., which apply to the resource even after it has been decrypted, be dealt with?

Within this model, it is not really possible to prevent a photo from being distributed, manipulated, etc., after decryption. Likewise, a video or soundtrack, once streamed, can easily be recorded and processed, even though such rights may not have been granted by a license. There is the chance of taking "a posteriori" actions (like tracing illegal distribution through embedding robust watermarks), but only after the damage has been done. also It is also possible to request a signature from everyone who uploads content (possibly modified), but this would simply transfer the issue of enforcement to yet another level: is it possible to prevent the upload of non-signed content?

Our conclusion is that only a trusted middleware running on trusted hardware platforms such as Trusted Platform Module (TPM)-based devices, allows fully-fledged enforcements of rights, capable of capturing all the possibilities REL has the ability of describing.

7.6.5 Fractals of Trust

The core concept behind the CONVERGENCE publish/subscribe model is the assignment of peers to fractals, according to the content they are contributing to the system. So, when a peer publishes some content within a fractal, the content is propagated to a number of other peers in the fractal. The same holds for subscription. In other words a fractal is a set of peers sharing some common interests. The concept and use of fractals in CONVERGENCE is fully described in Chap. 4.

In this section the CONVERGENCE approach is proposed to create a fractal based on criteria other than the users' interests (as expressed through the semantic descriptions of the resources), but also based on level of trust of a fractal. Even though the project has achieved some results in this domain, the scheme described below should be regarded as provisional. Further work would be required to deliver more conclusive outcomes.

7.6.5.1 Beyond Basics

As explained in Chap. 4, a users' specification of the fractal of interest may be expressed in terms of a semantic keyword, a flat keyword or an operation on these keywords. For example, it is possible to have the fractal MOVIE + ACTION for action movies. However, the fractals must not be necessarily bounded by the content circulated inside them; they can also be based on the properties of other peers, such as trust. This is explored in this section.

Specifying fractals based on trust can be seen as crucial not only for giving owners control over content, but also for balancing the need for performance against the need for protection. Assuming that the owner encrypts the VDI using some license and then propagates it blindly over some fractals, selected only by content, some receivers may not be able to handle the VDI. Conversely, there might be peers that would be allowed to access that content, but that will never receive it. The latter situation could occur for one of two reasons: (1) probabilistic protocols; and (2) limiting propagation depth. The technique proposed in CONVERGENCE, adding trust as one of the parameters to select fractals, provides another means of limiting the number of target peers, thus contributing to the scalability of the protocol.

7.6.5.2 Fractals of Trust

The question that arises at this point is how to define such a fractal. CONVER-GENCE envisages two solutions, one of which requires the assistance of the user, whilst the other is fully automated. In the current approach, it was considered the creation of a fractal to be dynamic and under the total control of the user. In parallel, a machine-based solution could be deployed, by automatically extracting descriptive metadata contained in the P-VDI. For example, if a user says that the content he/she is about to publish is MOVIEZ, then, the MOVIEZ fractal is created. It should always be kept in mind that fractals are just a virtual organization of peers and, hence, a peer may belong to more than one fractal.

Apart from the increased matching performance, another important property of a fractal of trust is that it allows a set of peers to create closed organizations, accessed by peers that correspond to certain criteria, keeping the symmetric nature of the system. This is achieved during the registration process, by protecting the fractal registry. At this point, it should be observed that, when a peer requests to enter the fractal, it communicates with another peer and asks for a part of the fractal registry. If this registry is protected, then the peer will be able to use it only if it has the required properties.

Another issue is how to manage/enforce rights on the fractal registry and, thus, how to support fractals of trust. Obviously, a solution based on one certificate per fractal does not scale. This is then a case in which Attribute Based Encryption can be valuable.

Consider that each peer has a set of attributes, perhaps stored in a smart card. Thus, the creator of the fractal who initially owns the registry, encrypts it using the access policy for this particular fractal of trust. From this point on, only peers that have attributes satisfying this access policy will be able to read the registry entries they need to discover their neighbors, so only they will be able to access the fractal.

7.7 Conclusions

The CONVERGENCE system aims to become a platform for the future Internet, enabling the secure distribution of resources and the realization of diversified and advanced business models in which resources transit through complex value chains. Digital resources are often covered by a license or contract, specifying a business relationship, including access and usage rights and conditions. Accordingly, the CONVERGENCE system has to offer a complete security infrastructure that enables the formulation and enforcement of complex and diversified access and usage rights and conditions.

Rights Expression Languages provide a means to declare those rights and conditions in a machine-readable manner. Most existing RELs have been developed for a specific application domain or do not aim at exerting subsequent enforcement of digital rights. Many lack formality in the definition of their elements and are thus unsuitable for automated operation.

The MPEG-21 part 5 standard can be seen as a highly flexible, general-purpose rights expression language, well suited to producing machine-readable assertions of rights. The ultimate aim of the standard is to kick-start technologies that correctly enforce access and usage rights over different types of digital resources and across a broad range of application domains.

Having already developed detailed application scenarios and considering the diversity of digital resources which potentially are subject to distribution and transactions within the CONVERGENCE system, the project has defined a CONVERGENCE governance and licensing scheme. This scheme is based on a number of specific own patterns and a sub-set of MPEG-21 part 5 REL elements. This scheme, whilst supporting all requirements already identified in terms of content protection and digital rights management, is sufficiently flexible to be extended to accommodate further needs. The way in which MPEG-21 REL elements are used has been thoroughly explored. It should thus be possible to make future extensions to the scheme at low cost.

The project has formulated detailed proposals for an extension of the already rich set of REL verbs, identifying complex business cases that the current REL is unable to accommodate easily.

The most interesting aspect of any real-world REL implementation, one that CONVERGENCE started to cover but that needs further effort and exploration, is to design novel integration patterns of advanced security technologies. These novel patterns should allow to enforce at least the fundamental aspects of the REL scheme at hand, taking into account the challenging trade-offs that arise when abandoning requirements for fully-trusted hardware and software, which may be deemed as impractical, in favor of trust chains and lightweight smart-cards.

References

J. Bethencourt, A. Sahai, and B. Waters. 2007. Ciphertext-Policy Attribute-Based Encryption. In Proceedings of the 2007 IEEE Symposium on Security and Privacy (SP '07). IEEE Computer Society, Washington, DC, USA, 321–334. DOI:10.1109/SP.2007.11 http://dx.doi.org/10.1109/ SP.2007.11.
ccREL Website, http://creativecommons.org/ns

Deliverable D4.1, CONVERGENCE project. Preliminary Definition of the Versatile Digital Item. http://www.ict-convergence.eu/deliverables, 2011.

IETF Network Working Group. An Internet Attribute Certificate Profile for Authorization. Request for Comments: 3281. April 2002. http://www.ietf.org/rfc/rfc3281.txt

ISO/IEC 21000-5:2004, International Standards Organization (ISO), JTC 1, SC 29, Information technology—Multimedia framework (MPEG-21)—Part 5: Rights Expression Language.

A. Lewko and B. Waters. 2011. Decentralizing attribute-based encryption. In Proceedings of the 30th Annual international conference on Theory and applications of cryptographic techniques: advances in cryptology (EUROCRYPT'11), Kenneth G. Paterson (Ed.). Springer-Verlag, Berlin, Heidelberg, 568–588.

ODRL Website http://odrl.net/

OMA DRM Release 2.2 Website, http://www.openmobilealliance.org/Technical/release_program/drm_v2_2.aspx

T. Pedersen. 1991. Non-Interactive and Information-Theoretic Secure Verifiable Secret Sharing. In Proceedings of the 11th Annual International Cryptology Conference on Advances in Cryptology (CRYPTO '91), Joan Feigenbaum (Ed.). Springer-Verlag, London, UK, 129–140.

M. Stefik, Letting Loose the Light: Igniting Commerce in Electronic Publication, In: M. STefik (ed.), Internet Dreams: Archetypes, Myths, and Metaphors (pp. 219–253), Cambridge (MA), MIT Press, 1996.

Chapter 8
Scenarios and Applications for CONVERGENCE

**Panagiotis Gkonis, Heinrich Hussmann, Alina Hang,
Evgeniya Ivanova, Sam Habibi Minelli, Daniel Sequeira,
Ileana-Catinca Bobric, Mihai Tanase, Valérie Legrand-Galarza
and Francis Lemaitre**

Abstract This chapter provides a technical overview and description of the four use cases and their corresponding applications that were developed in the framework of CONVERGENCE Project: Photos in the Cloud and Analyses on the Earth (under the responsibility of partner Alinari), Videos in the Cloud and Analyses on the Earth (under the responsibility of partner FMSH), Augmented Lecture Podcast (under the responsibility of partner LMU) and Smart Retailing (under the responsibility of partners WIPRO and UTI). Each subsection begins with a brief description of the considered use case, where the advantages of CONVERGENCE over current TCP/IP implementations are highlighted. The analysis goes on with the functional description of applications, alongside some implementation details and requirements. In the final sections of the chapter, the basic CONVERGENCE tools are described along with additional proposals for further exploitation of the considered use cases. The implementation of tools and applications was based on: (1) the functional requirements of use cases specified in CONVERGENCE Project Deliverable D.2.2; (2) the definition of the Versatile Digital Item (VDI) described in Chap. 5 and specified in CONVERGENCE Project Deliverable D.4.1; and (3) the CONVERGENCE middleware protocols and

P. Gkonis (✉)
Singular Logic SA, Nea Ionia, Greece
e-mail: ep5@singularlogic.eu

H. Hussmann · A. Hang · E. Ivanova
University of Munich (LMU), Munich, Germany

S. H. Minelli
Alinari 24 Ore (Alinari), Florence, Italy

D. Sequeira
Wipro Technologies (Wipro), Maia, Portugal

I.-C. Bobric · M. Tanase
UTI GRUP SA, bucharest, Romania

V. Legrand-Galarza · F. Lemaitre
Fondation Maison des Sciences de l'Homme (FMSH), Paris, France

F. Almeida et al. (eds.), *Enhancing the Internet with the CONVERGENCE System*,
Signals and Communication Technology, DOI: 10.1007/978-1-4471-5373-3_8,
© Springer-Verlag London 2014

technology engines described in Chap. 4 and fully specified in CONVERGENCE Project Deliverables D.5.3 and D6.3. Moreover, the impact of CONVERGENCE to the end users who evaluated all developed applications is highlighted as well.

8.1 Introduction

The considered use cases in the framework of CONVERGENCE play a fundamental role towards the deployment and evaluation of ICN networks. These use cases were carefully selected based on the needs of industrial (i.e. Alinari, WIPRO and UTI) and non–industrial partners (i.e. FMSH, LMU) in an effort to identify users' needs as the structure of today's Internet shifts towards content-centric architecture and implementation. In this framework, four use cases were adopted and implemented:

- Photos in the cloud and Analyses on the Earth (Alinari): This scenario is dedicated to the photographic market. Due to the large expansion of professional and semi-professional photographic production, we aimed at powering archive and photographers businesses making it easier to publish, retrieve and distribute authoritative content with ready-made licensing models;
- Videos on the Cloud and Analyses on the Earth: This scenario has been designed to improve the management of audiovisual archives and to exploit the potential of semantic techniques, when the same video resources are exploited several times in different contexts of use (analyses using different domain ontologies, posting on different video channels);
- Augmented Lecture Podcast: This scenario is based on a web-based lecture podcast application that enables students to revise lectures by watching video podcasts with synchronized slides. The use of CONVERGENCE demonstrates a new way to achieve podcasting and annotation functionalities that provides a common basis for collaboration and information exchange;
- Smart Retailing: This scenario is grounded on a VDI-enabled product supply chain, where VDIs are associated with products after they have been manufactured and released to the market. This new supply chain will enhance the retail market with a wide range of new services and operations, from which manufacturers, retailers and consumers will benefit.

8.2 Photos in the Cloud and Analyses on the Earth

8.2.1 Brief Description of Use Case

This use case tests whether CONVERGENCE can provide useful support for new business models in the photography business. The goal is to make it easier for photographers to create, describe and distribute their photos, improve access for users and generally facilitate the management of relevant services. All photographs are represented by VDIs, each containing:

1. The photo itself (or a link to the photo), possibly accompanied by a low-resolution version of the photo;
2. Metadata describing the photo defined by Alinari and including the date, time and place where the photo was taken, legal data on the author and owner of the photo, technical data about the photo (camera, lens, shutter time, aperture, ISO etc.), historical data about the site represented in the photo, and other metadata contributed by Alinari staff and third parties;
3. Licensing information representing the conditions at which photos can be licensed by a photographer to Alinari and by Alinari to an end user. Licensing information is represented using the CONVERGENCE REL (CONVERGENCE Project Deliverable D.4.2).

The dedicated applications provide Alinari staff with a user interface making it easy for them to create, publish, un-publish, describe and update the photos and VDIs. Free-lance photographers can access the Alinari service to create Resource VDIs and Publication VDIs. End-users can subscribe to photos (thus activating a search and collect agent) and possibly buy photos or related editorial materials using a local application connected to the Alinari server. Alinari provides both images and annotations to be channeled through the application. Furthermore, Alinari generates retail products and offers which may match the visual content.

The CONVERGENCE framework automatically prevents access to photos that have expired and performs garbage collection to purge expired copies from network storage. This feature is unique, considering that all competitors take advantage from creative content made available by the users and their loss of control along time on their rights. In fact, most users register to picture sharing platforms and sites with nicknames (most often they do not provide personal home addresses): they generate creative content and then while they forget their credentials or for any reason abandon to access their profile on such sites allowing their collection stay public with no statement about their wills. As a consequence, their creative contents are orphaned. CONVERGENCE is innovative in this sense as it allows the user to define the lifetime of the published content.

8.2.2 Bird's Eye View of the Deployment Framework

The actors of the Alinari application are the photographers and archive managers on one side and the final customers on the other side. The photographers and Alinari are image sellers. The customers can be editorial or educational clients who need images for their activities or publications.

The Photographers/Alinari can create and manage their personal photographic gallery; they can associate their photographs to the license models previously defined by the Archive Manager (on the basis of specific templates or model releases) while they can also publish the photographs in the marketplace.

Customers can execute searches in the entire marketplace; they can save the executed searches inside the marketplace (set of images); they can purchase by means of prepaid credits, the desired set of images and the corresponding use license.

8.2.3 Supported Applications

The Photographic Archive Management (PAM) application targets professional photographers and agencies wishing to manage their archives. Professional photographers currently manage their collections using proprietary systems developed over a number of years. The Alinari PAM application provides an alternative. The application, which runs on top of the CONVERGENCE middleware and network, helps users to store and manage their own photo-libraries and to make the pictures available through a (shared) market place.

As evidenced by the Fig. 8.1, the photographers (or the Archive Manager) can upload on the system images and annotate them. They can also decide if they wish to publish the image and also set appropriate rights regarding image access. On the other side, the end users can search for interested topics and be notified when matching occurs among their search query and other information that has been published.

CONVERGENCE also executes a matching with retail products (product-VDIs) and offers (offer VDIs) and the camera information embedded automatically in the EXIF record of the image. Thus, the user visualizes the searched image and eventually the offers related to the camera model and lenses with which the photographer has taken the picture (see Fig. 8.2).

Even if the PAM application addresses the professional market, it can also offer a wide opportunity for crowd generated content: users can upload their creations

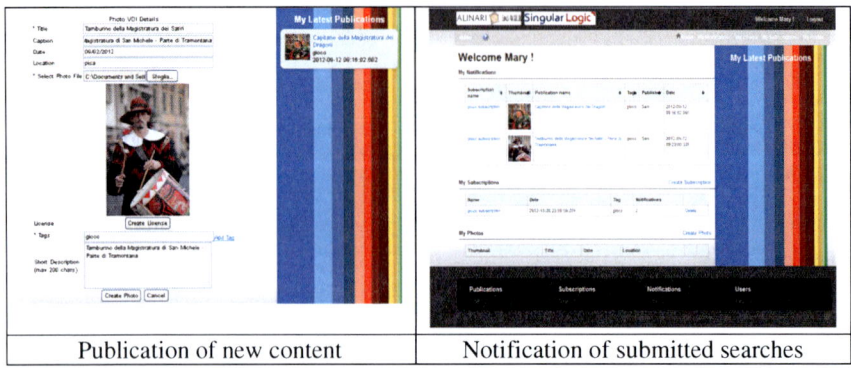

| Publication of new content | Notification of submitted searches |

Fig. 8.1 Publication of new content and notification in the Alinari application

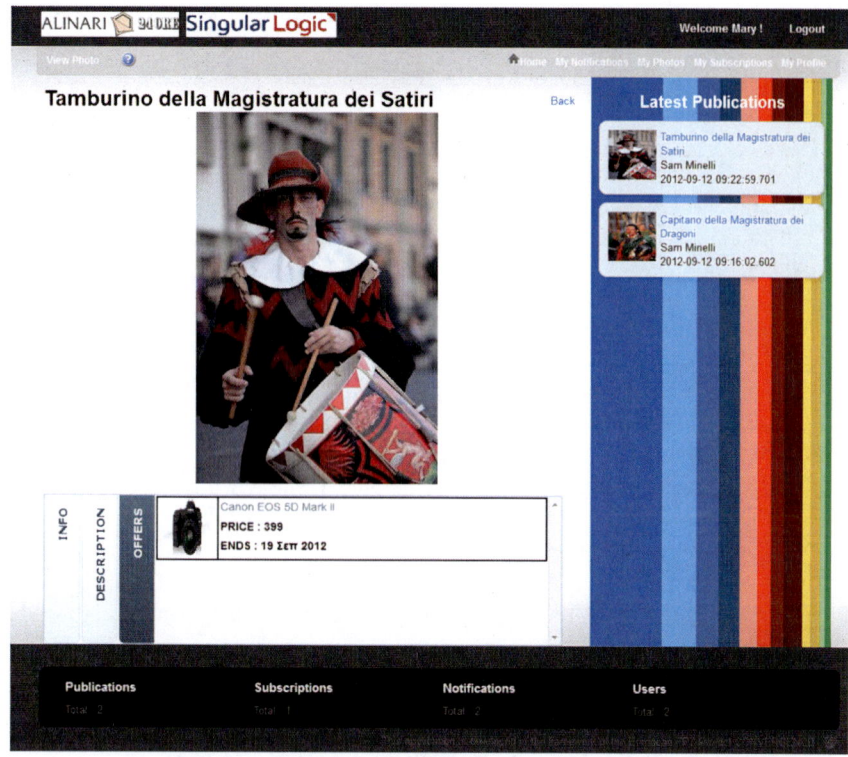

Fig. 8.2 Matching of the image and the related products

and share (publish) them. By the fact, the final PAM application has been addressed and designed to the social communities enabling them to look for interested collections and share published items in their own platforms (and thus attracting new users into PAM).

8.2.4 Functional Overview

The main activities of the applications in the framework of the PAM use case are depicted in the previous section. Users create, describe, apply licenses and publish VDIs that in turn can be subscribed to (searched for) by other users and potential clients. This section summarizes the functional requirements of the PAM application, which offers the following capabilities to the end users:

• Creation of Photo VDIs and storage in CONVERGENCE system;

- Enforcement of licenses in the photo with the help of CONVERGENCE REL;
- Creation of Publication VDIs and injection in the CONVERGENCE system;
- Subscription to photos under certain search criteria;
- Revocation of photos either after a predefined expiration date or on user demand;
- Registration to the application by different types of end users (i.e. free-lance photographers, Alinari staff, etc.);
- Authentication through a secure mechanism in order to avoid uploading of malicious content.

The most innovative functionalities are those related to the licensing (the user can define the visibility, access and use of the owned images) and to the retrieval (the photographer can publish and un-publish owned images; the client can activate some kind of temporally defined 'agents' who act on keys and provide the results of searches). A further innovative functionality implemented in CONVERGENCE is the matching of user generated content with retail products: as soon as the photographer or a client submits searches, CONVERGENCE performs keyword matching with existing retail products (Alinari provided 40,000 shop products in two languages ranging from art prints and collotypes to books, calendars and fine art prints): the user is shown the searched images and is advised on related offers.

8.3 Videos in the Cloud and Analyses on the Earth

8.3.1 Brief Description of Use Case

The "Videos in the Cloud and Analyses on the Earth" scenario illustrates how CONVERGENCE could provide new features for the management of audiovisual archives in the scientific and research areas. This use case is based on ESCoM-FMSH[1] platform of audiovisual digital libraries and archives, named ARA Campus[2] http://www.archivesaudiovisuelles.fr/EN/. The main goals and actions of this program, developing a scientific video library on-line, are:

---

[1] Equipe Sémiotique Cognitive et Nouveaux Medias (Cognitive semiotics and new media research team)—a research laboratory hosted in FMSH (Fondation Maison des Sciences de l'Homme), a French not-for profit scientific organization whose mission is to promote research in the social sciences and humanities.

[2] In 2001, ESCoM launched the ARA (Audiovisual Research Archives), a web video library dedicated to collect, archive, analyze and disseminate cultural and scientific heritages. Today the archives include more than 5800 h of videos in 16 languages, concerning scientific and cultural topics (such as social history, social and cultural anthropology, language studies and linguistics, archaeology, etc.).

- To collect and archive audiovisual corpora in order to constitute a multimedia portal of scientific heritage in social and human sciences;
- To analyze and describe audiovisual resources, with the elaboration of semiotic ontologies and analysis tools;
- To publish and disseminate these heritages on a web platform, making them available to specific target populations (researchers, educators, students etc.) through dedicated media channels;
- To investigate and develop the scientific means and technical tools enabling any institution or individual to create, manage, diffuse and share his own audiovisual resources in form of online libraries or archives.

In the context of Web video platform market for purposes of research, knowledge transmission or education, such as ESCOM-FMSH program, we point out 5 main needs:

- Compatibility and interoperability between the different existing audiovisual archive systems and technologies;
- Legal framework and juridical procedures through a scientific media and a licensing system allowing the protection of intellectual properties on videos and properly disseminate them;
- Autonomous and non-commercial services allowing people to autonomously produce, manage, diffuse, exploit and preserve their own audiovisual resources;
- Scientific uses and advanced functionalities, such as scientific annotation and indexing, facilitating the use of semiotics to new technologies and media;
- Online dissemination and remote access, allowing the sharing of audiovisual material and analyses from distant users communities.

Given the goals and characteristics of the ESCoM-FMSH scenario, the CONVERGENCE framework could provide a valuable tool to improve management of audiovisual data and to exploit the potential of semantic techniques, when the same video resources are exploited several times in different contexts of use. The integration of CONVERGENCE technology in ARA Campus will then answer to several central needs expressed by professional users, especially video and analysis owners:

- Using CONVERGENCE Licensing system to guarantee the respect of copyrights and intellectual property of content owners, the secure download and upload of resources, allowing flexible possibilities for sharing content;
- Using CONVERGENCE standard digital container VDI for the preservation of metadata, tracing any usage of the content;
- Using CONVERGENCE CDS to facilitate the definition of metadata and user search criteria;
- Using the CONVERGENCE subscription/publication mechanism to subscribe to content in a powerful way by the use of ontologies and semantic queries;
- Using CONVERGENCE event reporting to monitor the use of video resources and analyses, through notifications;

- Using CONET caching to reduce requirements for bandwidth and server resources and to optimize local delivery of video resources.

8.3.2 Bird's Eye View of the Deployment Framework

The ESCoM-FMSH scenario involves actors such as researchers, teachers, students and audiovisual professionals that belong to three main areas: research, education and culture. These actors are led up to play four different roles in the workshops developed by the ARA Campus for the management of videos, semantic resources, analyses and channels:

- Video Material Owners (VMOs) are people having full rights on video material. They publish encrypted video resources with licenses granting specific rights for using their material, and monitor the uses of their content (when a video is downloaded or analyzed by an Analyst, posted in a video channel, etc.);
- Analysts describe and analyze videos (i.e. discourse analysis, visual analysis, thematic analysis, etc.). They subscribe to videos, download video resources and publish video analyses with licenses. They monitor as well the uses of their analyses;
- Video Channel Owners (VCOs) are responsible of web video channels, proposing to browse video analyses and material following specific topics. They subscribe to video analyses of their interest by building semantic queries, then post analyses on their video channels;
- Video Channel Users (VCUs) subscribe to video channels and browse videos and analyses on the channels Fig. 8.3.

One important aspect of ESCoM-FMSH scenario is the need to enable VMOs to share their culturally sensitive material in a secured way that will ensure that their intellectual property, as well as the respect of filmed people's rights (indigenous

Fig. 8.3 Roles in ESCoM-FMSH application

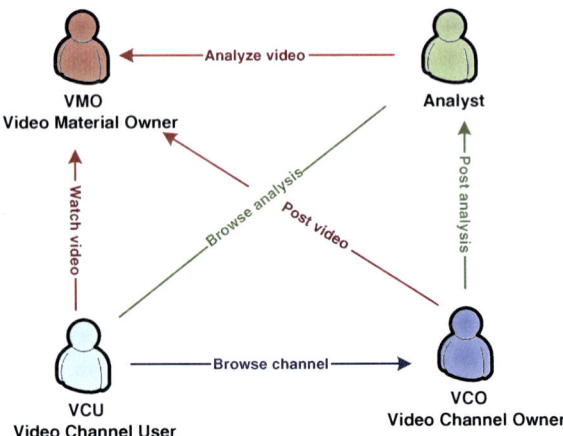

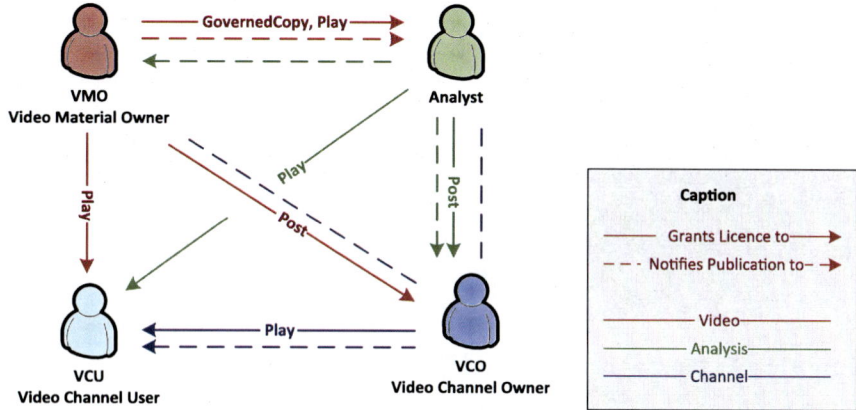

Fig. 8.4 Users interactions in ESCoM-FMSH application

people, researchers, etc.), are preserved. However, a same video can be used by different people (analyzed by a researcher, posted by a teacher, etc.), in different contexts of use (analyzed for specialists of a domain, analyzed for linguistics aspects, etc.). ESCoM-FMSH application answers to this need by:

- Enabling VMOs to manage the licenses for downloading, posting and watching their material, even when used by third parties (when analyses of their videos are posted in a video channel, for example);
- Enabling VMOs to keep notified every time their content is used, even by third parties: when their videos are referenced in an Analysis, when such an Analysis is posted in a channel, or even when their videos are watched in such a channel;
- Enabling Analysts and VCOs to manage the licenses and monitor of their own contents as well Fig. 8.4.

8.3.3 Supported Applications

The "Videos in the Cloud and Analyses on the Earth" Application provides a secured digital environment for the sharing and monitoring of video material, video analyses and video channels. The application offers several tools depending on the role of the authenticated user:

- Typically, a VMO upload a video file (stored encrypted in the network), then publish the video as a Video VDI with licenses (for downloading, posting and/or watching the video) and event report requests. He can monitor the uses of his video VDIs by the reception of notifications for the kinds of uses he requested to be notified;
- An Analyst usually receives notifications of new videos of his interest from his subscriptions. He provides the analysis of the video in a RDF format referencing

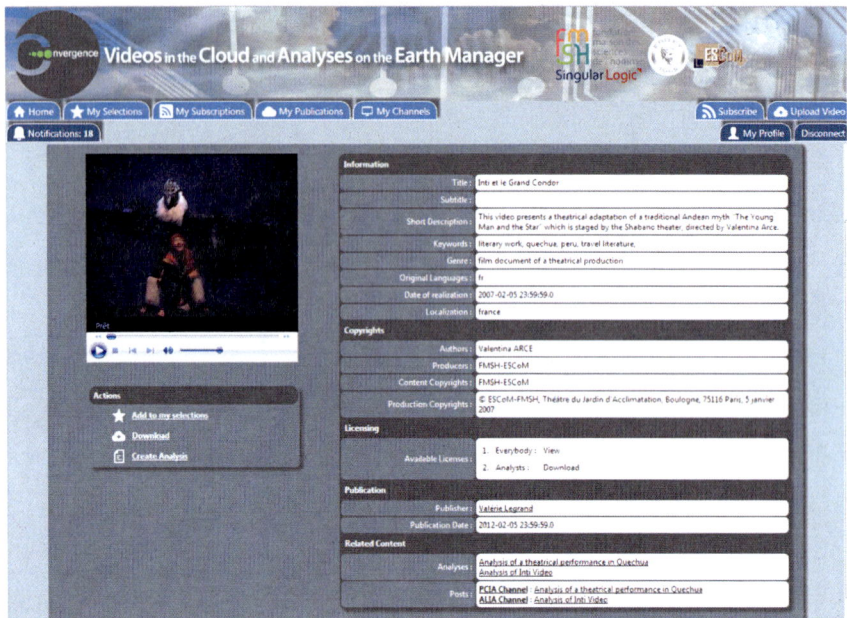

Fig. 8.5 Video browser

the relevant domain ontology stored in CDS Server. He publishes the RDF file as an Analysis VDI, with licenses and event report requests;

- The VCO receives notifications of new analyses of his interest from his subscriptions. These subscriptions consist of semantic queries which exploit the richness of the analyses metadata and domain ontologies (for example, subscribing to analyses of documentaries about traditional handcraft in South America). Then, the VCO can post the analysis in his channel, as a Publication VDI called Post VDI;
- VCUs subscribe to video channels and get notified of any new post in the channels they subscribe to Fig. 8.5.

8.3.4 Functional Overview

The ESCoM-FMSH application offers the following features to users:

- Upload and encryption of videos and analyses;
- Creation and publication of Video, Analysis and Post VDIs;
- Rights management with REL Licensing;
- Unpublication and digital forgetting;
- Publication/Subscription mechanisms supporting semantic queries;
- Support of different domain ontologies;

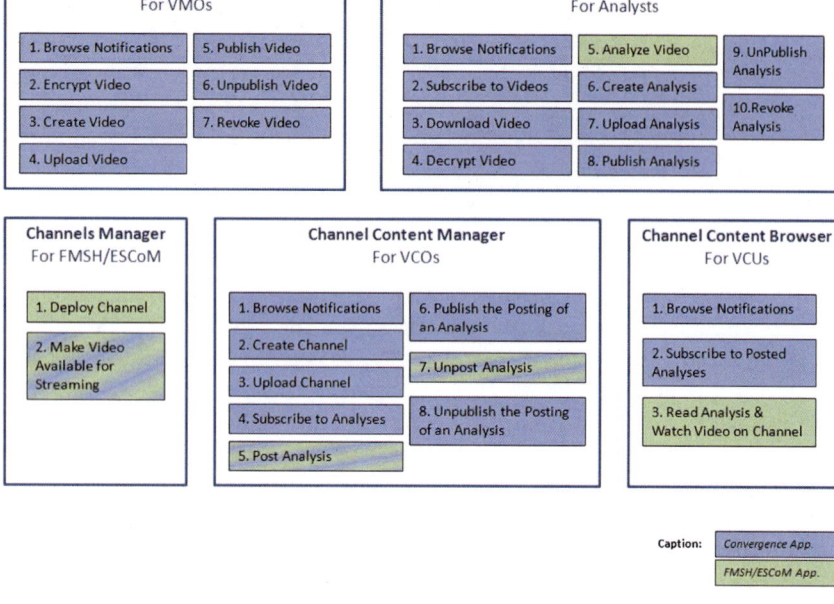

Fig. 8.6 Main features of ESCoM-FMSH application

- Video storage and delivery over CoNet;
- Content monitoring through notifications.

Key benefits of CONVERGENCE technology for users are the monitoring of the uses of their content (event report requests and notifications), as well as security enforcement (REL licensing, encryption, digital forgetting). This is more particularly valuable for Video Material Owners whose content can be indirectly used by Video Channel Owners (by posting analyses referencing their videos). The possibility to exploit the semantic metadata and ontologies for search, subscription and publication mechanisms is also an important aspect of the application Fig. 8.6.

8.4 Augmented Lecture Podcast (ALP)

8.4.1 Brief Description of Use Case

Traditional podcasts usually consist of audio/video streams and are sometimes accompanied by additional material (e.g. presentation slides). An innovative approach is to augment traditional podcasts with Web 2.0 features, allowing students to contribute their own content through annotations. The main goal of this idea is to engage students in discussions, enriching student-to-student communication as well as student-to-teacher communication. There already exists a variety of podcasting solutions that support this idea. For example, Moodle (Moodle n.d.) is a course management tool that supports lecturers in the creation of online courses. There also exist a variety of plugins for Moodle (e.g. podcasting plugins) to distribute podcasts and learning material. In fall 2012, MITx (MITx 2012) will start a new learning initiative where students can find a variety of courses from MIT embedded within a virtual learning community. Similar services can be found at other universities like the Open University (SocialLearn 2012) and LMU (Video Online n.d.).

The augmented lecture podcast scenario is based on the described ideas and employs a social learning environment where students watch video podcasts with synchronized presentation slides. Interactivity is achieved through private, semi-private and public annotations. The scenario only marginally shows new functionalities within social learning environments, but it demonstrates a new way of how these functionalities can be achieved using the CONVERGENCE framework.

We motivate the use of the CONVERGENCE framework due to the diversity of existing applications. Though all of them implement similar features, their basic infrastructure is completely different. That is, while one application uses Flash for implementation, others rely on PHP, JAVA, etc. This works fine as long as those standalone applications stay alone. However, we have to face some difficulties, when collaboration between applications is desired.

CONVERGENCE has the potential to overcome these issues. It offers a common underlying infrastructure that facilitates the collaboration, i.e. sharing content between applications. The main advantages of using CONVERGENCE are:

1. CONVERGENCE is a common standard. It creates interoperability and avoids platform lock-in;
2. CONVERGENCE maintains the relationship of all copies to the original source. When content is modified, automatic updates are made at the next adequate time. This ensures the quality of learning;
3. CONVERGENCE has an integrated notification mechanism. This enables asynchronous searches. Students do not miss the chance to find relevant content that is not published yet. They are notified as soon as new content is available;
4. CONVEGENCE ensures reliability of service. It allows content caching so that students can be offline to retrieve content (e.g. from students close to them).

8.4.2 Bird's Eye View of the Deployment Framework

The main actors of the augmented lecture podcast scenario are lecturers (content producers) and students (content consumers and producers). Their main activities are depicted in Fig. 8.7 and can be summarized as follows:

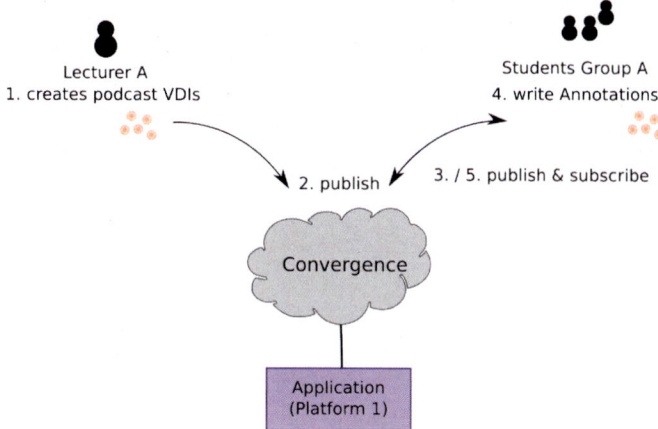

Fig. 8.7 Main activities of users within the augmented lecture podcast scenario

1. Lecturers create, store and publish lecture podcast VDI sequences, which are updated from time to time, when video or slides are modified. Lecturers revoke outdated lecture podcast VDIs. Lecturers can also retrieve statistical information about the use of the lecture podcast VDIs (statistical information is collected by the service provider);
2. Students search, subscribe, download and watch lecture podcasts. Students are also informed if a new lecture podcast or a new version of a podcast is made available by the service. Moreover, they can create and publish annotations to specific portions of the podcast using an augmented lecture podcast application which streams the podcast episode. Finally, they can receive notifications of published annotations and publish annotations to annotations as well as delete annotations.

Figure 8.7 depicts a single-application scenario. However, one can easily imagine an extension of this idea with additional applications. Due to the CONVERGENCE cloud, it is easy for applications that are running for example to another university to connect to CONVERGENCE and share their content (e.g. podcasts or annotations) with other existing applications. This is possible through the underlying common infrastructure that CONVERGENCE provides.

8.4.3 Supported Applications

The described activities are implemented in two CONVERGENCE-based applications: a podcast creator application for lecturers and an augmented lecture podcast service for students. Figure 8.8 shows two screenshots of the graphical user interface of both applications.

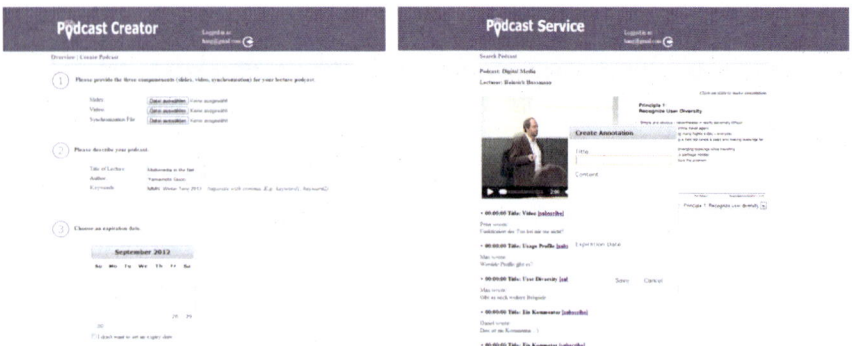

Fig. 8.8 Screenshot of two applications from the augmented lecture podcast scenario

Podcast creator: This application is used by lecturers to create, store and publish podcast components or podcasts by providing a variety of associated features. For example, lecturers can assemble different materials (i.e. presentations, videos and synchronization information) within VDIs and are guided stepwise through the creation process where they define the resources to be assembled, describe them and decide how long the resources are supposed to be available for search and retrieval by third parties. Once the information has been completed, the application publishes the VDI. Lecturers have the possibility to review, update and revoke (delete) their published VDI.

Augmented lecture podcast service: The podcast VDIs that have been published in the previously described application become available for the augmented lecture podcast service. The users of this service are students that search, watch and annotate podcasts. If they are particularly interested in certain content, they can take advantage of the CONVERGENCE subscribe functionality in order to get notified about newly published content, interesting annotations and other relevant information.

8.4.4 Functional Overview

The main activities of the applications in the framework of the ALP use case are depicted in the previous sections. The main idea is to base them on the CONVERGENCE framework. Users create, describe and publish VDIs that in turn can be subscribed to by persons of interest. With respect to our scenario, this means that lecturers will produce podcast VDIs sequences that students will subscribe to. This section summarizes the functional requirements of the two applications, which offers the following capabilities to the end users:

- Creation of podcast component VDIs/podcast VDIs and storage in the CONVERGENCE system;

- Definition of relationships between podcast component VDIs within a podcast VDI;
- Update of podcast component VDIs/podcast VDIs and revocation of outdated podcast component VDIs/Podcast VDIs;
- Creation of publication VDIs and injection in the CONVERGENCE system;
- Subscription to podcast component VDIs/podcast VDIs of interest
  - Augmentation of podcast VDIs with annotation VDIs;
- Selection of licenses for annotation VDIs (private, semi-private, public, anonymous)
  - Registration of users for both applications.

These application features/requirements enables user to experience the advantages of CONVERGENCE. In particular, users will get the idea of asynchronous searches, immediate updates and the work with always-up-to-date content. Content producers will experience the built-in protection of published content through enforcement of licenses and the easy management of published content. From a long-term perspective, where we assume a wide acceptance of CONVERGENCE and the existence of many CONVERGENCE-based applications available on the market, users will also experience the quality and richness of the provided content. Richness in the sense that a variety of different content will be available through the collaboration of different services that is facilitated by CONVERGENCE.

8.5 Smart Retailing (Real World Rrial 4)

8.5.1 Brief Description of Use Case

This use case describes a smart retailing supply chain for consumer electronic products. In this environment, CONVERGENCE provides users with a wide range of services and operations. The users and beneficiaries of this scenario can be divided in three distinct groups: Manufacturers, Retailers and Consumers. The Manufacturers create their own products VDIs and make them available for Retailers and Consumers by publishing the VDIs on the CONVERGENCE system. The Retailers subscribe to products from Manufactures and advertise the products they sell to Consumers with promotions, discounts, special offers, etc. Consumers subscribe to, search for and compare products through the CONVERGENCE network. Furthermore, additional services also improve the Consumers shopping experience allowing them to subscribe to information or specific data on products, receive notifications and sales events announcements about those products, and immediately look up for information on a particular product, accessing its VDI by scanning its barcode at the Retailers' stores.

In this retail scenario CONVERGENCE will aim to provide new services mostly to Retailers and Consumers, which will generate new and additional revenues for the entire retail industry. These new services produced by CON-VERGENCE offer a considerable number of significant innovations:

- A standard digital container (VDI) for product information, digitally signed by its Manufacturer;
- Retailer and Consumer access to up-to-date product information distributed via the CONVERGENCE Network;
- Integration between CONVERGENCE technology and existing retail tools and frameworks;
- Facilitated migration of product information from VDIs to Retailers' systems and databases;
- Improvements on in-store product management (product recalls, warranty verification, control, fraud and piracy prevention, etc.);
- CONVERGENCE-enabled POS machines, providing digital receipts and warranties to Consumers;
- CONVERGENCE-enabled retail applications that operate on top of CON-VERGENCE middleware.

8.5.2 Bird's Eye View of the Deployment Framework

From a high-level perspective this scenario involves all the following activities, listed below by users:

1. Manufacturers

   a. Create, store, advertise and publish Product Type VDIs containing reliable information such as the product name, description, physical characteristics, technical features and resources (product photos, user manuals, etc.).

2. Retailers

   a. Subscribe to Product Type VDIs created and published by Manufacturers;
   b. Receive notifications about all kinds of updates and warnings on Product Type VDIs they have subscribed to;
   c. Create and publish Retailers' Product Type VDIs, augmenting Manufacturers' Product Type VDIs with information about Retailer promotions, discounts, special offers, sales information, etc.;
   d. Create Product Instance VDIs when selling products to Consumers at the POS machine, adding new data such as consumer information, serial number and warranty details to Product Type VDIs;
   e. Visualize statistics about subscriptions made by consumers, divided into categories of products; monitor the customers' preferences and habits;
   f. Add customizable attributes to an offer made for a category of products, defining different properties of the products included in the offer;
   g. Create offers and promotions for an entire category of products, not only for a single product in particular.

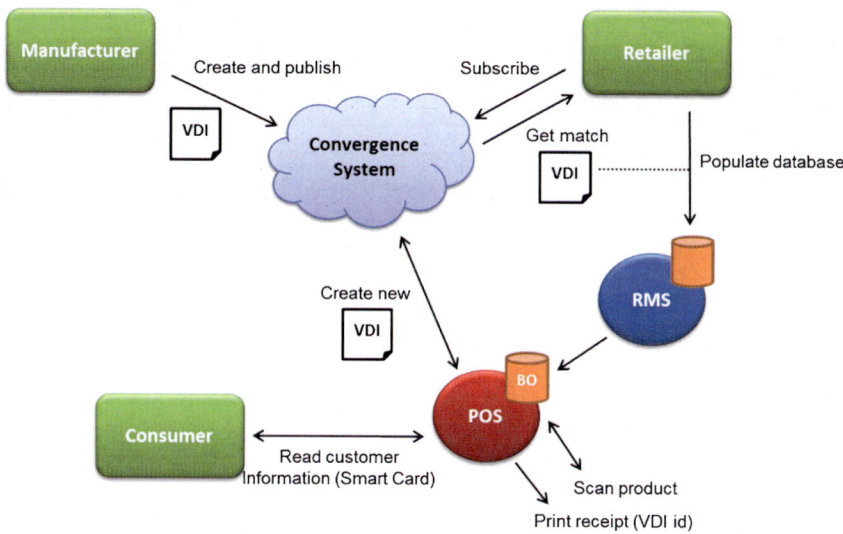

Fig. 8.9 Global view of the smart retailing scenario

3. Consumers

 a. Browse and subscribe to all kinds of product VDIs;
 b. Receive notifications about all kinds of updates and warnings on product VDIs they have subscribed to;
 c. Access and consult reliable information about products;
 d. Subscribe to an entire category of products and receive notifications about any product belonging to the category they subscribed;
 e. Pick attributes from a list and attach them to a subscription made on a category of products, defining particular properties that the products included in notifications should have (e.g. color, sizes).

The global view of the smart retailing scenario is shown in Fig. 8.9, where different types of VDIs related to the product supply chain are generated and published in CONVERGENCE.

8.5.3 Supported Applications

The described activities above were implemented in three CONVERGENCE-enabled applications: two "Smart Retailing" web applications (Figs. 8.10 and 8.11, WIPRO and UTI) to create, publish, subscribe and browse product VDIs and a Point-of-Sale application (Fig. 8.10) by WIPRO for every sale made by the retailers to consumers. Some screenshots of these applications are shown below.

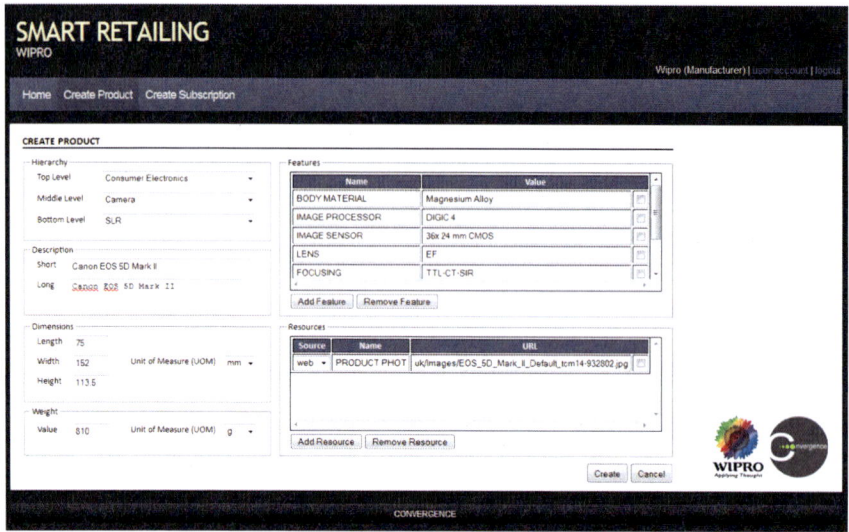

Fig. 8.10 Smart retailing web application "create product" form

In Fig. 8.10 it is shown a screenshot of the form in WIPRO's web application that is used to create, store and advertise a Product Type VDI containing product information (metadata) and resources (files), such as photos or user manuals. This create product form appears once the user clicks on the "Create Product" option of the web application.

Figure 8.11 displays the screen of UTI's web application from the retailer perspective. The user has the possibility to view details about the offers, the matches and the subscriptions on the same screen. By clicking on the number listed in the "Matches no." field, the retailer can see the subscriptions that matched his offer.

In Fig. 8.12 it is shown a screenshot of the retailers' POS machine Oracle application. When selling items (products) to consumers, the sales agent begins with the "Sell Item Screen" where all the transactions begin and end. This screen always shows all the items involved in the current transaction between the retailer and the consumer.

8.5.4 Functional Overview

The development of all the applications mentioned above involved the following requirements:

- Registration by different types of users: manufacturers, retailers and consumers;
- Authentication of different types of users through a secure mechanism;

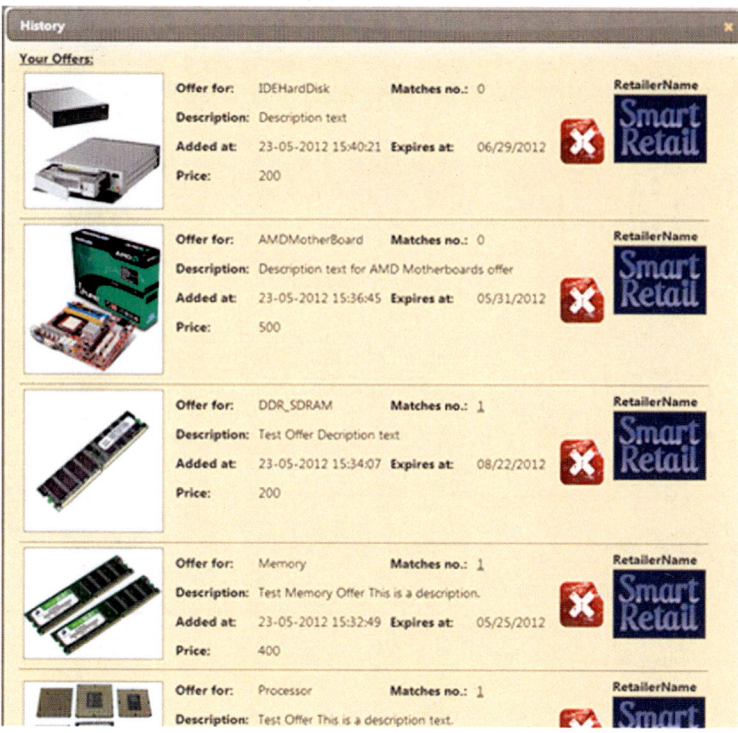

Fig. 8.11 UTI smart retailing application—visualize the offers, the matches and the subscriptions

- Creation of Product VDIs and storage in CONVERGENCE system;
- Enforcement of licenses in the Product VDI data and resources with the help of REL;
- Creation of Publication VDIs and injection in the CONVERGENCE system;
- Subscription to products under particular keyword(s) criteria;
- Display of personalized notifications related to the subscribed products;
- Revocation of Publication and Subscription VDIs after a predefined expiration date.

These requirements, used to develop all the retail applications for this scenario, are the prime foundation to develop any CONVERGENCE-enable application, regarding user experience, applications functionalities and their interaction with the CONVERGENCE middleware. In this context all of these requirements mentioned above were tested throughout all the three phases of the end users real world trials, not only for these end users to experience all the advantages of CONVERGENCE, but also for the developers to perceive their validity and sustainability facing real world problems. In general, users that participated in trials were able to get familiar with the product VDI concept by creating their own

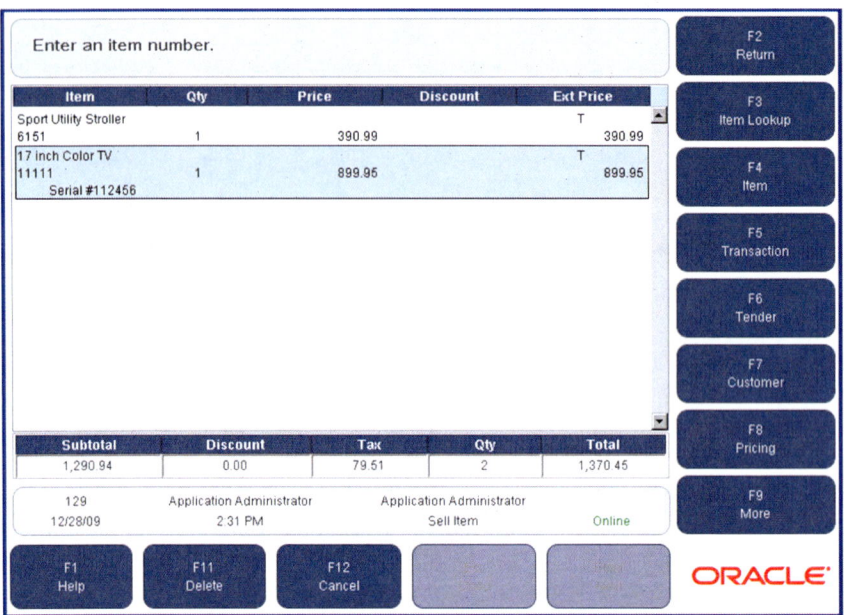

Fig. 8.12 Point-of-sale application screenshot of shell item screen

products VDIs and understanding the publish/subscribe mechanism implemented on CONVERGENCE. They also became aware of the notifications system related to their own products subscriptions and apprehended the way all the content flows in CONVERGENCE.

8.6 Convergence Tools and Deployment of Applications

In the previous paragraphs the CONVERGENCE use cases were presented while their decomposition in supported applications was described as well. Moving from specifications towards implementation, the goal of the consortium was to identify similar operations among applications that could be deployed once and reused many times by different applications: This task had two goals:

- On one hand to facilitate the work of the application developers;
- On the other hand to create units of reusable code that in future can be used in order to develop additional applications based on more complex use cases.

The overall architecture of tools and applications is depicted in Fig. 8.13. As it can be observed, the Application level provides the interface between users and CONVERGENCE and is the top level of CONVERGENCE from the user point of view. The Application level is split into two sub-levels: a Tools sub-level and a

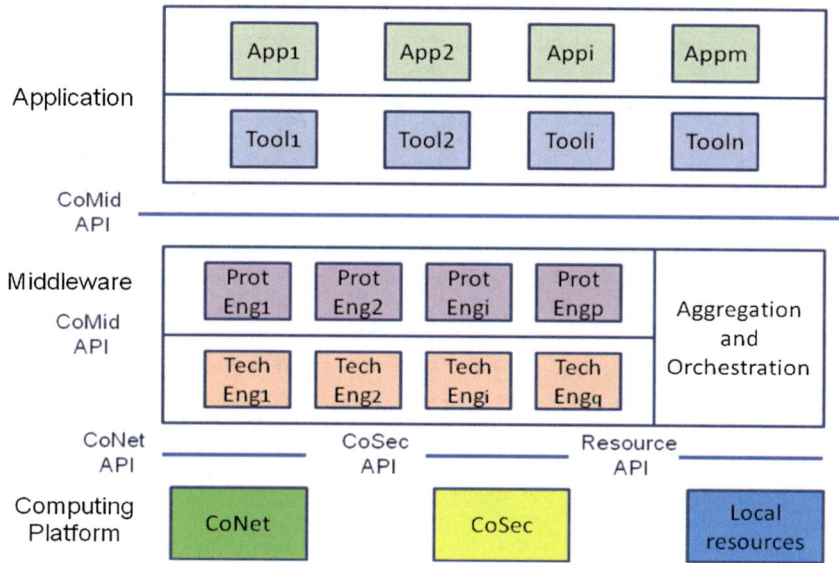

Fig. 8.13 Tools and applications architecture

User Applications sub-level. The Tools sub-level contains functional components that combine a subset of CoMid functionalities and can be reused by many applications but not all of them. In each case, the user interface to the tool is provided by the application, with different applications presenting different interfaces.

Following the analyses of the previous paragraphs, it becomes apparent that the applications of the CONVERGENCE use cases can be decomposed in a set of five fundamental operations:

- User registration to the application: All application should be in position to provide to their end users a simple but secure mechanism to register and authenticate their selves;
- Creation of Application VDIs (Resource VDIs): Different types of VDIs should be supported, such as Photo, Video, Podcast, Product and Offer VDIs. This tool should also be in position to provide to the end user the capability of inserting among others licenses inside the VDI regarding its usage as well as methods to digitally sign it;
- Publish of VDIs in CoNet: Users should be in position to advertise their resources in CONVERGENCE cloud by injecting VDIs consisting of a subset of information related to the original Resource VDI. As in Resource VDIs, the insertion of appropriate licenses should be supported as well;
- Subscription to VDIs: Users should be able to express their interest over a digital resource by injecting their search query in CONVERGENCE. This is performed by the creation of Subscription VDIs. Matching is typically performed among

the metadata of a Publication and a Subscription VDI provided that certain conditions defined with the help of CONVERGENCE REL are met;

- Revocation of VDIs: Removal of VDIs on demand should be supported in all types of applications. When such an event takes place, hosting peers of this VDI should be informed as well.

Therefore, in the framework of CONVERGNECE five fundamental tools were developed:

- **User registration:** The first step in each scenario is for the user to register with an identity provider, who provides her with a certificate and a personal smart card, which stores an automatically generated key pair and the certificate. The user also receives a unique identifier. After registering with the identity provider, the user can register with a service provider, using a pseudonym (if this is allowed by the application). After registering with the Identity Provider and the Service Provider, the user can authenticate herself, either using a username and password or the smart card. Usually the content provider (i.e. Alinari or FMSH) act as a service provider as well, while the identity provider is a third party trusted authority;
- **Resource VDI:** This tool allows a user to create and parse a Resource VDI (R-VDI) and provides methods to add and extract metadata, licenses, resource references and links to other VDIs. Additional methods allow users to package the VDI with resources, to sign and to store the VDI. These functionalities are supported by various middleware engines, the most important of which is the VDI TE. The metadata element of the R-VDI is created and parsed by the Metadata TE. Therefore, application users can create photo, video, podcast or product VDIs, add metadata and enforce licensing conditions;
- **Publish VDI:** This tool allows a user who has successfully created a VDI to create, inject and parse a Publication VDI (P-VDI), providing methods to add and parse metadata, licenses, and event report requests. Other methods allow users to inject a P-VDI into the overlay, to add a digital signature and to store it. These functionalities involve various middleware engines, the most important being the VDI TE. Metadata elements are handled by the Metadata TE. The Publish VDI tools shares several methods with the Resource VDI Tool. Most of these methods are supported by the same middleware engines in both tools. Therefore, the R-VDIs that were created in the previous step can now be advertised to the CONVERGECE system under appropriate license conditions, thus allowing other CONVERGENCE users to be informed on potential updates and/or purchase digital resources;
- **Subscribe VDI:** This tool allows a user to create and parse a Subscription VDI (S-VDI). The tool provides methods for adding, and extracting conditions, licenses, event report requests. Additional methods allow the user to inject the S-VDI into the semantic overlay, to add a digital signature or to store it. These functionalities are supported by various middleware engines, the most important being the VDI TE. The creation and parsing of metadata is handled by the Metadata TE. Subscription queries are formulated using the MPQF TE;

- **Revoke VDI:** This tool allows CONVERGENCE users to revoke a P-VDI or an S-VDI from CoNet or CoMid. The tool uses CoNet TE for CoNet revocation and the Overlay TE to propagate the revocation request to peers.

8.7 Additional Fields of Exploitation

Once the development of the CONVERGENCE applications was finalized and validated through the CONVERGENCE trials, the consortium identified additional fields of exploitation of these applications. This final use case integrates three individual CONVERGENCE applications: The Alinari Photographic Archive Management Application (PAM) and the two retail applications from UTI and WIPRO. The goal is, on the one hand, to demonstrate interoperability among applications, on the other hand to show how CONVERGENCE could support additional use cases, not described here.

The use case (see Fig. 8.14) involves two user profiles, one for the content creator (a photographer) and one for the content consumer. The business stakeholders are Alinari (which sells image rights and fine art books and printed material), WIPRO and UTI (both technology providers for the retail industry). Once the photographer has created photographic images (including automatically generated camera information, and published her creations through the PAM interface) CONVERGENCE associates the photo with the Product VDIs for her

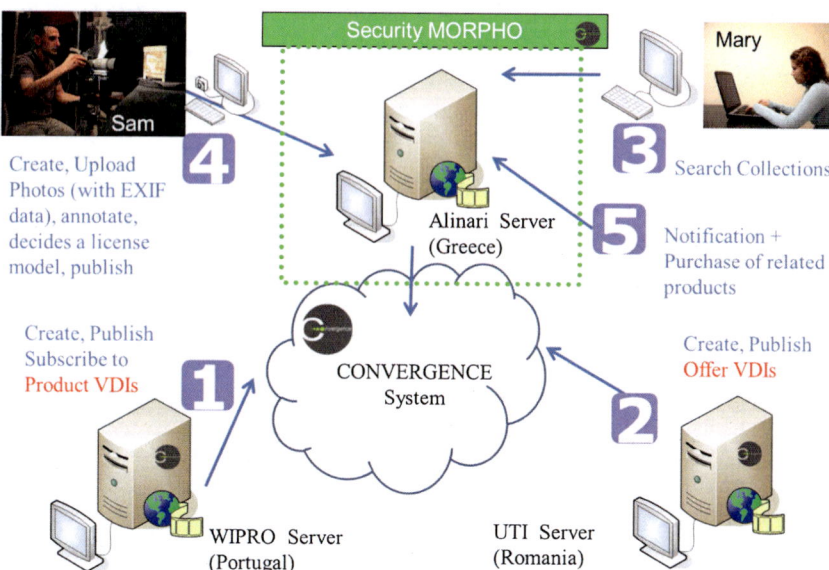

Fig. 8.14 Integrated Alinari/WIPRO/UTI application

camera and lenses, created with the WIPRO application and with special offers to buy the equipment (created with the UTI application). When the content consumer retrieves images through the PAM, she receives similar information.

Users registered to the Alinari application are able to:

- Create, Annotate, License and Publish photos (photographers);
- Submit queries, personalizing the period to which the query refers (photographers and content consumers);
- Purchase products related to digital photos, such as cameras and lenses (photographers and content consumers).

To support this scenario, the application automatically matches EXIF information incorporated in the photo with product VDIs, for cameras, lenses etc. Other trusted subjects (e.g. UTI), can then use this information to generate offer VDIs, which can then be visualized through the PAM application.

In brief the application provides the consumer with information that could lead her to make new purchases—in this case, cameras and lenses, in the future books from the photographer, Alinari books and posters, etc.

For this system to work, a manufacturer has to publish for example a camera product VDI in the CONVERGENCE system, before bringing a new camera to the market. The camera product VDI is published in a specific fractal that the Manufacturer shares with her certified Retailers. Once the camera product VDI is published, all retailers that are allowed to subscribe to the fractal (all authorized retailers) will have the new product in their catalogue and will be able to make special offers to customers wishing to purchase the product. To create such an offer, a retailer defines an offer VDI linked to the Manufacturer's product VDI. The offer VDI will contain specific fields for the offer (e.g. price, period of validity for the offer, picture of the promotional material). This VDI is published in a new fractal. After this step, every subscriber to this fractal will be notified about the new offer. The notification will allow the user to browse the details and go to a link where she can purchase the product.

8.8 The Convergence Trials

All the applications described above were thoroughly tested through the CONVERGENCE trials. For this reason, each scenario leader recruited several external users that were invited to evaluate the developed applications in terms of new supported functionalities and eventually express their opinion over the proposed CONVERGENCE solution. The trials were organized in three distinct rounds, where users' comments and feedback from the first two rounds were taken into consideration by the developers' team in order to enhance certain aspects of the involved applications. As it was extracted from users' reports, CONVERGENCE provided sufficient solutions to a number of issues related to the management and protection as well as search and retrieval of digital content:

- Better control over the data users share (e.g. setting expiration dates), as well as over the information that other people can find and access about them. This is due to the explicit ownership rights that CONVERGENCE proposes (control over digital resources through the enforcement of licenses as well as maintenance of anonymity when for example publishing content);
- Content is always kept up-to-date without the need to worry about outdated information or about multiple copies of the same content. In this context, CONVERGENCE offers the "future search" functionality to its end users. This practically means that matches among a subscription and a publication VDI may also occur at a future time since there can be publications made after the specific subscription. This feature is crucial for the ALP application for example where students and lecturers want to be always kept up to date when searching for educational material. Moreover, in the WIPRO and UTI application retailers and consumers that have subscribed to particular products may receive instant notifications when for example a new version of the product is released along with relevant documentation. Additionally, this feature is enforced with the compatibility with semantic search (powered by CDS Server), allowing the user to search for very specific content and to get notified of publications matching exactly with their interests (experimented in FMSH scenario where rich semantic metadata of video analyses are exploited);
- Efficient traceability of content published in the network, enabled by the Event Reports Requests and Notifications features of CONVERGENCE. This point has been noticed by the users involved in the trials as a key advantage, seen as a powerful technology for collaborative tools. This is also very useful for contents which can be re-used by different people, and/or exploited in different contexts of use (particularly in FMSH scenario where a same video can be analysed by several analysts, and where Video Material Owner need to keep informed when analyses of their material are posted in video channels);
- Interoperability and matching among different types of VDIs generated in CONVERGENCE, such as Photo VDIs, Product VDIs and Offer VDIs in the integrated Alinari/WIPRO/UTI application for example. Therefore, users can be notified on interesting products and potentially purchase them when for example they are viewing a photo VDI in the PAM application.

8.9 Conclusions

In this chapter, the CONVERGENCE use cases along with the corresponding applications were presented and analyzed. The analysis of the applications was decomposed in five fundamental operations: Subscription and registration to applications, creation of application resource VDIs, publish of VDIs, subscription to VDIs as well as revocation of VDIs. Therefore, the corresponding tools based on these operations were implemented, thus facilitating the deployment and testing of all applications.

As validated by the CONVERGENCE trials, the concept of VDI provides solutions to a number of issues related to the management of digital content: dynamic update and revocation of content, digital resource protection, event report requests and notifications, enforcement of rights over digital resources, interoperability among different types of digital content, etc.

However, as it was presented in this chapter, the road to additional fields of exploitation remains open. In particular, the first integrated application made possible the purchase of specific products such as cameras and lenses, related to photographic digital content. The presented use case shows how different types of VDIs (i.e. photo, product, offer VDIs) can be accessed and processed by one single integrated application; thus facilitating ICN applications interoperability.

The outcomes of this integration are important for realistic ICN networks, where it is expected that different types of digital media content will be exchanged in the network. Therefore, as search and retrieval of content is of primary importance, ICN applications should be in position to deal with different types of content and present it to the end users.

In this context, we believe that the interoperability demonstrated in the integrated application framework is relevant not just for business applications but also for education. Taking the FMSH and LMU scenarios as an example, it would be possible, for instance, for two institutions to share video material annotated by their respective users' viewers, while using different software solutions and pursuing different educational goals.

References

MITx: MIT's new online learning initiative, (2012), Retrieved November 6, 2012, from http://mitx.mit.edu/.

Moodle. (n.d.), Retrieved November 20, 2012, from https://moodle.org/.

SocialLearn. (2012), Retrieved November 6, 2012, from http://sociallearn.open.ac.uk/public.

Video Online Unterrichtsmitschau 2.0. (n.d.), Retrieved November 6, 2012, from http://videoonline.edu.lmu.de/.

Chapter 9
Business Models and Exploitation

**Sam Habibi Minelli, Andrea de Polo, Giuseppe Tropea,
Panagiotis Gkonis, Alina Hang, Mihai Tanase, Daniel Sequeira,
Francis Lemaitre and Riccardo Chiariglione**

Abstract This chapter provides the CONVERGENCE business Models for the commercial and non-commercial exploitation of CONVERGENCE applications and technology. The chapter collects the studies of feasibility and implications of alternative exploitation strategies. We identified competing products and services, and evidenced the market risks and threats.

9.1 Introduction

The current CONVERGENCE consortium comprises different categories of organization (industrial companies, SMEs, universities, research institutions) with different interests, strategies and business cultures. Some of these organizations

S. Habibi Minelli (✉) · A. de Polo
Alinari 24 Ore (Alinari), Florence, Italy
e-mail: sam.minelli@gmail.com

G. Tropea
Electronic Engineering Department, University of Rome "Tor Vergata", Rome, Italy

P. Gkonis
Singular Logic (SIL), Athens, Greece

A. Hang
University of Munich, Media Informatics Group, Munich, Germany

M. Tanase
UTI Group, Szczecin, Romania

D. Sequeira
WIPRO, Moreira, Portugal

F. Lemaitre
Fondation Maison des Sciences de L'Homme (FMSH), Paris, France

R. Chiariglione
CEDEO, Villar Dora, Italy

F. Almeida et al. (eds.), *Enhancing the Internet with the CONVERGENCE System*,
Signals and Communication Technology, DOI: 10.1007/978-1-4471-5373-3_9,
© Springer-Verlag London 2014

have access to significant financial resources. For others, even relatively small investments are problematic. For the consortium nature and structure it is unlikely that the current partners could formulate a single business strategy capable of meeting their very different requirements. Thus, in this chapter we will present commercial and non-commercial business models covering the nature of partners and in some cases we will specify more in detail the local business model.

CONVERGENCE is a complex system, which can serve as the basis for many different products and services, targeting different user populations and communities. Work in the project so far has led to:

- A complete specification and reference implementation of the CONVERGENCE middleware (CoMID) and network (CoNET), including features very likely to become part of MPEG standards;
- Prototype smartcards (hardware and software) providing advanced functionality to protect the privacy and security of CONVERGENCE users and to protect CONVERGENCE CONTENT;
- Four prototype applications, dedicated respectively to stock photography, video sharing in the non-profit market, distance education and retailing.

This does not mean however that the partners are ready to exploit the results of their joint research. At the technical level, this will require a significant effort to industrialize the current prototypes and to improve the design of the interfaces and documentation provided to end-users and developers. In marketing terms, the partners will need to establish a CONVERGENCE brand, and to convince potential customers of the value of the technology. At the organizational support, it will be necessary to find solutions that guarantee the future of the CONVERGENCE framework, providing on-going maintenance, development and technical support. All this requires planning. In this report, therefore, we report the current state of the partners' plans.

The CONVERGENCE Consortium includes commercial companies (Alinari, CEDEO, Morpho, SIL, UTI, WIPRO, XIW), non-profit organizations (FMSH) as well as Universities and research institutions (CNIT, ICCS, INESC-Porto, LMU). Obviously these organizations have different interests and different strategies for exploitation. The commercial organizations in the Consortium have developed detailed plans to exploit the technologies and applications they have helped to develop; FMSH has developed a clear plan to incorporate CONVERGENCE in the not-for-profit services it provides to the French academic community. The academic partners, on the other hand are more concerned with the development of CONVERGENCE as a public good, and with the exploitation of the technology in their research and teaching. Fortunately, as we will see the interests of these two very different categories of partner coincide. The academic partners will guarantee the future development and maintenance of the CONVERGENCE framework, using it in their teaching and research and promote CONVERGENCE to the scientific world and to standards organizations; the commercial partners will develop applications and technologies with clear prospects for commercial exploitation.

9.1.1 Outputs of the CONVERGENCE Project

A fully deployed CONVERGENCE system would consist of:

- One or more, potentially interconnected networks of CONVERGENCE nodes providing CONVERGENCE publish-subscribe functionality to CONVERGENCE applications;
- Applications that access and exploit this functionality via the CONVERGENCE middleware (CoMid).

At the time of writing, the CONVERGENCE partners have developed and tested prototype implementations of the CoMid and of its underlying functionality, prototypes of related security technologies (smartcards) and several prototype applications (for professional photography, for the sharing of video resources in a non-profit context, for "lecture podcasts" for "smart retailing", for video streaming). These prototypes are intended to provide a proof of concept for the CONVERGENCE approach and are not yet ready for deployment in an operational environment.

From this brief overview it is possible to identify the minimal requirements for the exploitation of CONVERGENCE results:

- Industrialization and maintenance of the CoMid including all required technology and protocol engines;
- Deployment and operation of one or more CONVERGENCE networks;
- Development and deployment of one or more CONVERGENCE applications.

If the exploitation is to be successful it will also be necessary to meet other requirements, in particular in terms of marketing: the ultimate value of CONVERGENCE will depend on the number of users and applications it can attract.

Meeting these requirements would need significant investment by the CONVERGENCE partners, and, perhaps new legal and organizational arrangements. Obviously such investment can only be justified if it brings adequate returns. In this report, therefore, we examine potential solutions and analyse their costs and benefits for the CONVERGENCE partners. In doing so, we bear in mind that CONVERGENCE is a mixed consortium including organizations (universities, commercial companies, non-profit organizations) with very different goals, perspectives and markets.

9.2 E-Commerce Photographs and Fine Art Production

This business scenario has been contributed by Alinari, the oldest photo archive in the world with over 5.5 M historical and modern images from around the world. The way Alinari does business is by licensing rights to use its images for commercial, personal and educational purposes. The company is also famous for its production of personalized high quality photographic books and special editions for important companies. Last but not least, Alinari sells unique reproductions of early seventeenth century drawings with a unique continuous tone process called collotype. The company, with 25 employees and photographic archive partners around the world, is recognized as a leading fine art photo archive and content provider in Europe and abroad. It has already digitalized 350,000 images and indexed them in a multilingual database. New images are added on a daily basis.

9.2.1 Description of Business Scenario

Alinari needs to establish a truly credible and convincing business position in the publishing market, for photography and fine art. The current market trends evidence increasing demand of pictures but at lower prices with the bad effect of copyright infringements and inadequate image right protection. Therefore, the company is interested in the prospect of creating an online e-commerce web site focused on real-time licensing of photographs, and online sales of its publications and collotype drawings. To date, however, those ideas have not being fully implemented and it is not possible to distribute pictures under established licensing models (traditional professional photographers are used to decide the licensing models on the basis of the client requests: let it be for advertisement, for marketing, for documenting or video broadcasting, etc.).

The two main difficulties are Alinari's need to control the DRMS and content licensing workflow, and its need to manage the online catalog, all in real time. In this setting, CONVERGENCE can provide the quality, credibility, and efficiency Alinari requires. The planned trial in Alinari will thus demonstrate the integration of CONVERGENCE in the Alinari online e-commerce portal, where it will be used to sell and distribute rights for Alinari digital photographs, fine art photographic books and collotype reproductions of drawings. More specifically VDI's will play a critical role in synchronizing the Alinari picture database (currently based on Microsoft SQL Server and IIS DB) and the Alinari merchandising catalogue with the online shopping portal. VDIs will be used in various steps of the digitization process, including image digitalization, content indexing (including "region of interest" picture descriptions), semantic and metadata tagging, injection of GPS information, publishing, licensing and selling. As stated earlier, one of the most important advantages of VDIs for Alinari and its picture agency partners is that they can provide a shorter and monitored distribution model for visual products with immediate sell (currently, many Alinari's clients, wait some days to acquire the desired images).

9.2.2 Users

Alinari is currently exploring possibilities for commercial exploitation of the Photo Archive Management (PAM) tool developed in collaboration with SIL and UTI (adding Alinari's editorial offers through the application).

The PAM application targets professional photographers and agencies wishing to manage their archives. Professional photographers currently manage their collections using proprietary systems developed over a number of years. The Alinari PAM application provides an alternative.

The application, which runs on top of the CONVERGENCE middleware and network, helps users to store and manage their own photo-libraries and to make the pictures available through a (shared) market place. With the application:

1. Photographers can

   a. Create and manage their gallery;
   b. Associate their photographs to the license models previously defined by the Archive Manager;
   c. Publish the photographs in the marketplace.

2. Customers can

   a. Execute searches in the entire marketplace;
   b. Save the executed searches inside the marketplace (set of images);
   c. Purchase by means of prepaid credits, the desired set of images and the corresponding use license (*).

3. Alinari's Archive manager can (*)

   a. Manage the use licenses and assign a pricing model to them;
   b. Manage the market place services;
   c. Monitor the orders and the image flow (download processes);
   d. Offer related editorial products matched by CONVERGENCE to the user subscriptions.

9.2.3 The Market, Products and Services

In what follows, we limit our detailed estimates to the market we know (Italy) and offer only rough estimates of the European market, considering any additional revenue flows as a "bonus". We estimate that Italy alone has $\sim 8,000$ professional photographers. From our knowledge of the market, we estimate that each of these photographers is willing to spend up to Eur 150 per year for the kind of services offered by the PAM app. On this basis, we estimate that the overall Italian market is worth Eur ~ 1.2 million per year while the European market is worth more than

Eur ~ 10 million per year. If we also include assistants and semi-professional photographers we can increase our target population in Italy to more than 16,000 units (and a corresponding market of Eur ~ 2.4 million). The total size of the European market amounts to between ~ 10 M Eur and ~ 70 M Eur per year. Alinari can probably address 5–10 % of this market.

Services to internal team in Alinari CONVERGENCE technology supports a different paradigm to manage content and make it available to the clients: the licensing can be fixed in advance and be linked to the content; clients and content providers can be made more autonomous and thus involve less costs for Alinari. Internal team in Alinari can provide packaged contents (thematically chosen) to clients and advertise collateral products by combining user searches to matched books and retail products.

Services to photographers professional photographers and agencies would pay a basic fee for using the service, plus royalties on sales of licenses. Browsing would be free.

Services to end users final users will be able to acquire content and licenses directly with no human interaction in the Archive and save searches (submissions) defining the period of interest for pictures. This would also enable the Archive's photographer to plan photographic campaigns.

9.2.4 Table of Transactions

Considering that Alinari has already applied some of the knowledge it acquired during the CONVERGENCE project, using it to speed up its delivery procedures and updates. In particular, Alinari has used VDIs concept model in a recent project with the Municipality of Milan (2010–2011), in which it provided the municipality with more than 4,000 images for the local tourism web site: http://www.turismo.milano.it. In this project, we used VDIs concept model as a means of packaging images (in different cuts and resolutions), metadata, licensing information and machine-to-machine information in a single information container. Alinari used VDIs again during the national photographic competition www.sognomilano.alinari.it. During the competition, Alinari received several thousand pictures from photographers, which it had to process very rapidly, providing a license for each for each photo before sending it to the evaluators and to the Municipality of Milan. In brief, VDIs have enabled Alinari to implement new workflows providing higher quality faster delivery, and generally enhancing Alinari's ability to compete at lower market prices. From these experiences Alinari planned a 5 year cash flow estimation and the financial indicators as shown in the Table 9.1.

Table 9.1 Cash flow and financial indicators for Alinari scenario

	Y1	Y2	Y3	Y4	Y5
Revenues (worst case)	€58.380,00	€145.102,00	€414.802,00	€616.004,00	€963.606,00
Costs	€121.000,00	€128.000,00	€162.500,00	€160.700,00	€172.400,00
Revenues—costs	**−€62.620,00**	**€17.102,00**	**€252.302,00**	**€455.304,00**	**€791.206,00**
Required external investment	€62.620,00	€0,00	€0,00	€0,00	€0,00
Internal rate of return	196 %				
Net present value at 10 %	€949.020,27				
Break-even	Y2				

9.2.5 Business Flow

Alinari's Business Flow combines internal and external factors. Internally Alinari is aiming at reducing costs and improving the production flow. Externally Alinari is aiming to better collect and resell professional photographers and external represented archives products. Furthermore Alinari aims at selling other editorial products from CONVERGENCE ability to match user search and e-commerce products and offers made by Alinari (Fig. 9.1).

9.2.6 Comparison with Traditional Implementations

Existing services for access to large photographic libraries have nothing equivalent to VDI synchronization and authenticity solutions. Most existing services are offered by commercial companies or communities and are based on proprietary technologies. This means that content owners have no control over the technology. This can create concern about security, trust and privacy issues. Last but not least, the management, update and harvesting of those solutions can be tedious and complex, especially for users who need to update and synchronize large volumes of content.

9.2.7 Business Models

B2B (business): In this case, content is sold to professional licensees (editors, magazines, publishers). Services and searching tools are personalized according to the needs of the customer and to the business contracts signed.

The business model for this scenario is to offer the CONVERGENCE solution to photographic partners and agencies that already work with Alinari. Alinari would allow these partners to sell their content through PAM interface on a

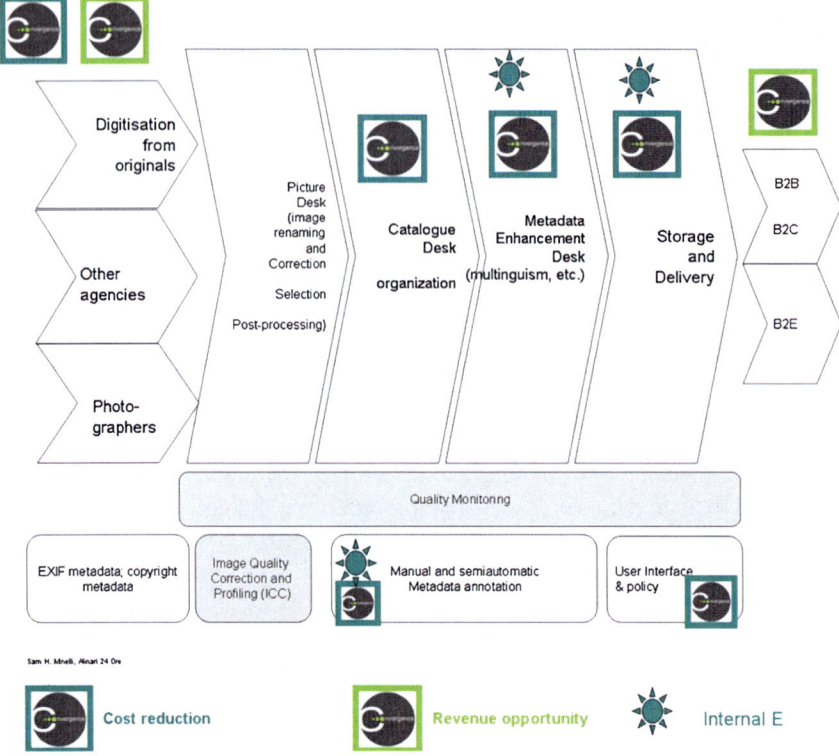

Fig. 9.1 Business flow in Alinari business scenario

revenue share model. Thus Alinari would distribute and sell third parties collections and share the revenues with them.

B2E (education): Alinari would offer a free access with restricted features and an annual subscription for flat access to content. Additional collections would be available on request under a pay per upgrade basis.

B2C (consumer): This solution would offer free access with restricted features, limited access to search functionality and the matching with Alinari's e-commerce product will generate the revenue stream.

9.3 Semantic and Secured Digital Environment for Non Profit Audiovisual Web Channels

In 2002, FMSH's R&D lab in digital audiovisual medias ESCoM (*Equipe Sémiotique Cognitive et Nouveaux Médias*) lab launched the "Audiovisual Research Archive" Program (ARA). ARA has a two-fold mission:

1. To produce a digital audiovisual online library that documents the scientific and cultural heritage in social and human sciences;
2. To perform research and development in the use and exploitation of audiovisual content.

Almost ten years later, the ARA Program offers:

1. A collection of almost 6,000 h of online video resources accessible through a main web portal and several specialized portals;
2. A digital environment for working with audiovisual corpora called ASW[1] Studio (Studio ASA in French). The environment provides tools for segmenting, describing, annotating, indexing and publishing audiovisual corpora as well as meta-linguistic resources—the ASW ontology—for building the libraries of description models necessary for the segmentation, analysis, and publishing of audiovisual materials;
3. A technical infrastructure for storing, streaming, using and preserving audiovisual data (essentially a group of servers maintained by the FMSH Computer Resources Department).

At the beginning of 2012, ESCoM started an exploratory R&D project (*ARA Campus*) whose goal is to provide an appropriate environment for individual academics and academic groups working with audiovisual resources. ARA Campus is intended to:

1. Become a video portal offering access to video resources provided by individual academics and academic groups;
2. Provide these actors with the tools they need to create and manage their own archives, working and publishing their video collections, using and/or reusing their own collections and/or those of third parties.

9.3.1 Description of Business Scenario

ARA Campus is a technical and scientific platform that in principle enables any academic or academic group to:

1. Create and manage an online audiovisual archive (a "video channel" in the sense of YouTube);
2. Upload audiovisual resources to the archive;
3. Delete and change audiovisual resources in the archive;

---

[1] « ASW » stands for « Audiovisuel Semiotic Workshop » ASWwas developed thanks to an ANR (Agence Nationale de la Recherche) funded R&D project called « ASA-SHS » (Atelier de Sémiotique Audiovisuelle pour l'analyse de corpus audiovisuals en SHS): http://www.asa-shs.fr/ (also: http://asashs.hypotheses.org/78).

4. Develop and manage ontologies (concepts, description models and thesaurus) on the basis of a common ontology provided by ARA Campus;
5. Define publishing formats for audiovisual resources on the basis of a "library" of models provided by ARA Campus;
6. Segment, analyse, publish and republish audiovisual resources;
7. Incorporate videos provided by third parties (other researchers, other institutions, etc.);
8. Moderate the segmentation, description and republication of resources by third parties;
9. Moderate revision of ontology—changes proposed by third parties using the archive as a platform for storage and dissemination;
10. Moderate third party requests to use records from to the archive for commercial or non-commercial purposes;
11. Trace all uses of audiovisual resources belonging to the archive;
12. Distribute parts of archive material over social media, tracing the way they are used.

ARA Campus acts as a content provider for:

1. Third party videos;
2. Third party video archives;
3. Third party video descriptions and video publications (in the form of interactive video books, thematic folders, video-lexica, etc.).

9.3.2 Users

As mentioned earlier, the main "market" for ARA Campus is the *public academic world*, primarily in France. This comprises:

1. Individual academics: researchers, teachers, doctorate and post-doc students;
2. Collective and institutional actors: research projects and programs, research labs, research networks, associations.

More specifically, the market comprises:

1. Individual or institutional actors who already own audiovisual data and who want to digitize the data for purposes of preservation and dissemination;
2. Individual or institutional actors who want to create their own audiovisual archive systems for research, educational or cultural purposes;
3. Individual or institutional actors who want to use or reuse already existing digital resources for their own specific activities (in research, in education, in documentation, etc.) (Fig. 9.2).

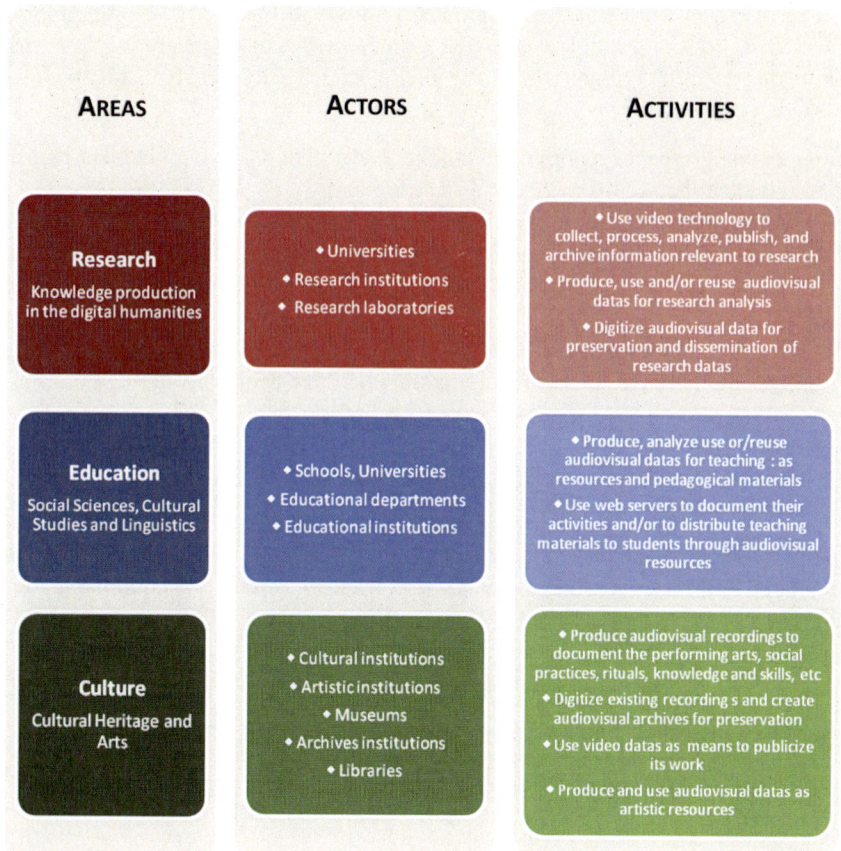

AREAS	ACTORS	ACTIVITIES
Research Knowledge production in the digital humanities	◆ Universities ◆ Research institutions ◆ Research laboratories	◆ Use video technology to collect, process, analyze, publish, and archive information relevant to research ◆ Produce, use and/or reuse audiovisual datas for research analysis ◆ Digitize audiovisual data for preservation and dissemination of research datas
Education Social Sciences, Cultural Studies and Linguistics	◆ Schools, Universities ◆ Educational departments ◆ Educational institutions	◆ Produce, analyze use or/reuse audiovisual datas for teaching : as resources and pedagogical materials ◆ Use web servers to document their activities and/or to distribute teaching materials to students through audiovisual resources
Culture Cultural Heritage and Arts	◆ Cultural institutions ◆ Artistic institutions ◆ Museums ◆ Archives institutions ◆ Libraries	◆ Produce audiovisual recordings to document the performing arts, social practices, rituals, knowledge and skills, etc ◆ Digitize existing recording s and create audiovisual archives for preservation ◆ Use video datas as means to publicize its work ◆ Produce and use audiovisual datas as artistic resources

Fig. 9.2 Target market actors

9.3.3 Services

ARA Campus consists of several services (or "workshops") for the management of videos, semantic resources, analyses and channels. In addition, ARA Campus will host a cloud providing the remote services needed by these workshops, including storage, querying of metadata, video streaming, etc. In this setting, ARA Campus will exploit specific CONVERGENCE services to provide:

1. Licensing;
2. Secure download and upload of resources;
3. Subscription/publication mechanisms;
4. Event reports requests/notifications.

An overview vision of ARA Campus is shown in Fig. 9.3. The integration of CONVERGENCE technology in ARA Campus will satisfy several central needs expressed by professional users. The use of CONVERGENCE technology will make it possible to:

1. Guarantee respect for copyright and the intellectual property of content owners;
2. Guarantee the secure transaction of resources;
3. Offer flexible possibilities for sharing content, based on specific licenses;
4. Trace the way content is used, making sure that it is not altered or used in a context that could affect the image of the those concerned (the video owner, people appearing in the film, the analyst, etc.);
5. Subscribe to content using ontologies and semantic queries;
6. Send automatic notifications to authors whenever content is used.

9.3.4 Comparison with Traditional Implementations

This scenario shows several significant improvements over traditional implementations. It introduces new features, and makes it possible to combine features that were previously available only as stand-alone services. The main improvements can be summarized as follows:

1. Audiovisual web channel administrators can implement semantic descriptions and publishing templates based on semantic properties without depending on the help of a technical developer. Anyone wishing to create an audiovisual web channel can set up his/her own video description and publishing templates;

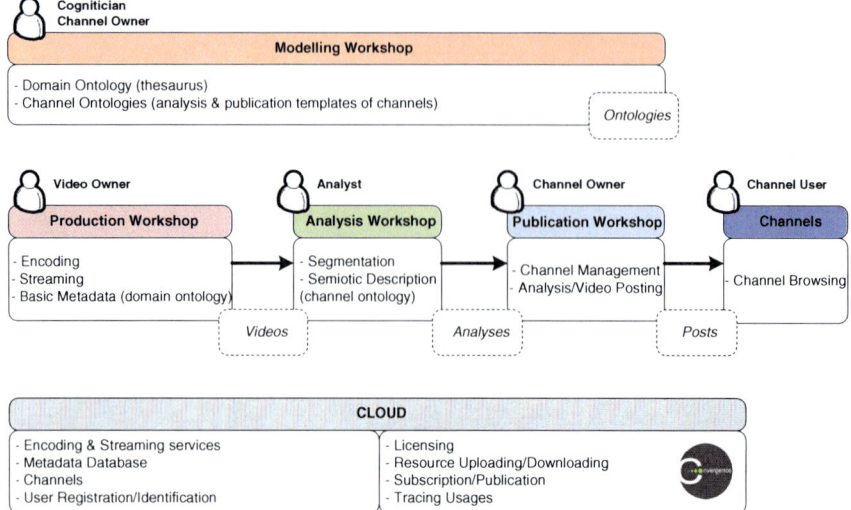

Fig. 9.3 ARA campus services

2. Audiovisual web channel administrators can set up properties which are common to a video corpus—properties such as rights management (licensing, user rights, copyrights, etc.) and video trans-coding properties. The improvement here is that these properties are described (VDI format) and used (CONVERGENCE network) in the same way as video descriptions, video annotations or video web portals;
3. Digital forgetting VDI adds strong support for digital forgetting that data providers (video provider, channel editor, video analyst...) can trust;
4. Data providers are able to control and update both:

   a. information linked to content they provide (e.g. video notices and copyrights);
   b. the list of users who are authorized to exploit, and re-use video resources and the way they use or re-use it.

In CONVERGENCE, developers only have to deal with one class of object (i.e. VDIs), in whose key features (e.g. licensing rights) have a standard representation.

9.3.5 Business Models

This offer a significant number of business possibilities such as:

1. Licensing of VDIs to public institutions (such as museums, archives or libraries) and social (virtual or "real") communities intending to create an online audiovisual library (for instance, in cultural heritage): the CoApp Provider sells or cedes free licenses to a VDI "service" for the registration of collected videos, the management of IPR, the "watermarking" of stored videos, the trans-coding of the stored videos in different streaming formats and online publishing of stored videos;
2. Licensing of VDIs for use in education: the CoApp Provider sells or cedes free licenses for a VDI "service" to educational institutions (schools, universities, ...) and individuals working in education (teachers, tutors, ...) for the creation of video channels, the constitution of online personal video corpora, the (virtual) segmentation of videos for teaching purposes objectives and the use of specific domain ontologies to enhance the pedagogical interest and relevancy of parts of previously segmented videos;
3. Licensing of VDIs for translation activities: the CoApp Provider sells or cedes free licenses of a VDI "service" to individuals and/or institutions working in the audiovisual translation market, enabling these actors to open monolingual video resources to the intrinsically multilingual digital knowledge market;
4. Licensing of VDIs for public communication and valorization activities: the CoApp Provider sells or cedes free licenses of a VDI "service" to individuals

or (public or private) institutions, enabling them to disseminate and valorize their (cultural, historical, scientific, …) heritage in form of short video trailers, online publicity, inserts in electronic messages, etc.

9.4 Pay Video Services Supply

Since 2009 CEDEO has been working on the development of PDAT (Platform for Digital Asset Trading), a technology platform based on the first edition of ISO/IEC 23006-2 MPEG Extensible Middleware (MXM). PDAT is centred on a server and supports a number of clients. The platform provides the following basic functionality (non-exhaustive list):

1. Input/output of a video;
2. Description of a video;
3. Creation of content (audio + description);
4. Posting of content;
5. Negotiation of content;
6. Licensing of a video;
7. Reporting of events;
8. Payment for a video;
9. Search for a video;
10. Streaming of a video.

Since 2011, CEDEO has developed a first line of business based on PDAT. This is CEDEO's WimTV service (http://wim.tv/). Since 2012 CEDEO has also developed a second line of business based on the licensing of customized PDAT solutions.

CEDEO is interested in migrating PDAT and its two business lines to support the CONVERGENCE technology. Such a move would offer a number of benefits to the company. In what follows we will limit our discussion to the use of CONVERGENCE security technologies.

9.4.1 Description of Business Scenario

Even though live streaming technologies have been around for some time, they have seldom been implemented as pay services and have tended to serve mainly niche markets. However, this situation is changing. With downstream bit rates of several Mbit/s now routinely available, new opportunities are arising to offer coverage of events (religious, social, sport, cultural etc.) whose audiences are too small to justify the use of broadcast channels (DTT, satellite and cable).

It is difficult to give a precise estimate of the size of the market. To get a first idea, consider the C-league in Italian football. The league stages 85 matches per year. Assuming 6,000 viewers per match (a conservative estimate) and a price of €2.5 per match, a service showing C-league matches could earn about Eur 1,300,000/year. Extrapolating this result to the European Union level, we obtain an estimate of Eur $\sim$ 10 M. We expect that as the new service expands, other smaller football leagues will also want to benefit. In the end, total revenues could be of the order of Eur $\sim$ 100 M. If other categories of customers also used the service revenues could be even higher

To exploit these opportunities commercially, providers need adequate protection for streamed content and a flexible online payments scheme. These are the services CEDEO would seek to provide.

We assume two entities: Event Organizer and Event Reseller (in some cases they could be the same organization). The former is the holder of the rights to the event; the latter is the organization that actually distributes the event. WimTV offers live distribution and split payment services to both and could extend the service with the current WimTV offering of on-demand video streaming distribution and payment services (Fig. 9.4).

CEDEO plans to produce new C++ implementations of its video clients incorporating the following new features:

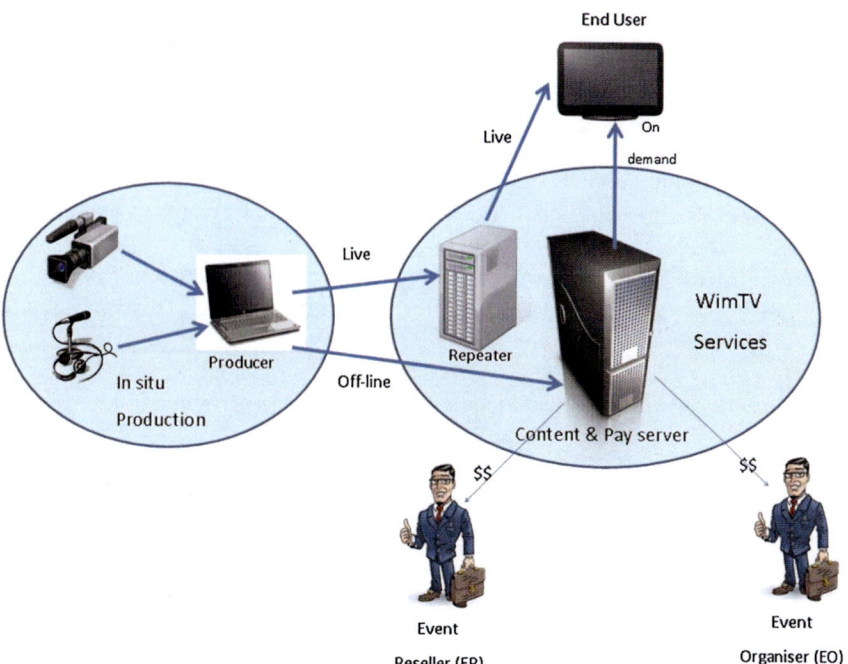

Fig. 9.4 WimTV business model

1. Licenses expressing user rights to video (e.g. the right to download);
2. Enhanced security using the CONVERGENCE Security TE and hardware-based security for user identification (e.g. user group identification), user authentication, and content protection.

These technologies will allow CEDEO to extend its current business—essentially based on "on-demand" services—to "live" services.

9.4.2 Users

Users of WimTV live streaming service come from a wide range of entities. Basically, any person, company, association, organization and even institutions are potential adopters of WimTV solution.

Below, a list of candidate users is provided, together with a motivation explaining why they would be interested.

- Sports: football teams, as well as basket, volley, and many other sports can proficiently offer live streaming to their supporters;
- Theatres: many theatres today are already equipped for filming (for archiving purposes) the most important shows that they host. It would be very easy—and effortless—for them to live stream these shows to the interested audience. Theatres could share the deriving revenues with the performers, thus generating a new revenue flow for both parties;
- Teachers, schools: schools and independent teachers can use live streaming functionality for e-learning sessions;
- Exhibitions: live streaming is a powerful way to increase the interest of people in products, and services presented in exhibitions. Organizers can stream the most important events (such as workshops, conferences, speeches) that take place during the exhibition;
- Music festivals and independent music bands: concerts are one of the most promising types of event to exploit live streaming services. This way they can "sell" virtual tickets to remote audience interested in the performance;
- Institutions and public bodies: town councils and other important moments of the life of a town can be streamed online, so as to be participated also by those who live too far or cannot reach the physical event.

9.4.3 Services

WimTV live streaming services make ti possible to video shoot events and transmit them online, on the web and on other devices like smartphones, tablets and SmartTV.

Table 9.2 Cash flow projections and financial indicators for WimTV scenario

	Y1	Y2	Y3
Cash flow			
Revenues	70	937	6,586
Costs	296	425	636
Revenues—costs	−226	512	5,950
Accumulated cash flow	−224	289	6,241
Financial indicators			
Internal rate of return	11 %		
Net present value at 10 %	4,596		
Break-even at month	22		

Live events can be offered for free, pay per view, or subscription based. For instance, theatres or other event organizers who want to stream large number of live shows, can offer to their audience a bundle. By paying a subscription fee, users are granted access to all events included in the bundle (Table 9.2).

There are also important non-financial benefits. In general terms, the new service will increase the global perceived value of CEDEO in terms of:

1. The range of services the company can provide;
2. The range of customer needs it can meet;
3. The expertise of CEDEO personnel.

This is important for revenues from consultancy, an area in which CEDEO has been active for many years.

9.4.4 Business Flow

CEDEO's main objective with the deployment of live streaming services is to increase its revenues by offering its customers an innovative way to generate a profit by exploiting live events.

Live streaming is not an isolated service. WimTV also offers many others functionalities that serve the needs of those who work with video content.

Once transmitted, live events can be published and offered as video on demand in a owned web tv channel, always with the possibility of creating revenues through pay per view. Event organizers and event resellers can create a wide catalogue of all the events they have offered live, and make them available to users at any time from their web tvs.

9.4.5 Comparison with Traditional Implementations

WimTV is not the only platform offering live streaming functionalities. You-Stream, Livestream, Ustream among the others, offer something similar.

The main difference between WimTV and other competitors is that WimTV give its customers very flexible conditions for setting up a business. When they create a live event, customers not only can specify the price of the virtual ticket to view, but also can involve other parties in the deal (like music festival organizers and the bands performing in it).

Generally, when more parties collaborate to offer services or content, and generate revenues, the main hurdle they face is about trust. Do they trust each other? Who is—among different parties—the one appointed to cash payments? How can it give evidence to other parties about the total income they got from the shared business? How can other parties be sure about the correctness of the revenue share? Often, these difficulties are a barrier for entities who want to collaborate. They need lawyers, expensive contracts, which at the end can turn the joint business in a loss of money.

WimTV overtakes all this since, when an economical agreement is reached among parties, all payments are split according to the percentages they set in their WimTV dashboard. No lawyers and other costs are needed, in order to make parties trust one another.

The introduction of secure split payment service, and the possibility to allow different entities to operate in the same deal (the transmission of a live event) is the most important difference between WimTV and other solutions.

9.5 Smart Product Purchase

We explored also other business models not limiting ourselves to produce local business solutions for partners of CONVERGENCE. Among all, we can envisage potential markets in the coming 'smart products'. In fact, we imagine that as soon as products will become more actively informative (as example by addressing the counteract market), by enabling products to be 'smarter', then CONVERGENCE system will have attractive potentials for investors.

9.5.1 Description of Business Scenario

In this scenario, the CoMid provider certifies individual manufacturing companies that wish to advertise their products to the public. Each VDI contains a broad set of information relating to a product. Customers may subscribe to this service in order to shop more easily. In this case, a customer connects to CoNet through her mobile device and enters the name of a product and the manufacturer. The network responds with the associated VDI as well as with VDIs of similar products. The VDI contains information such as the indicative price of the product, as well as locations where it can be found. Alternatively, this information can be provided by entering the product's unique identifier (such as barcode). Stores can also subscribe to the service. In this way, they can publish VDIs containing the list of available products.

Consumer organizations may also subscribe to the service, so as to provide consumers with additional information and guarantee that products available to the public are safe. CoApps will support the appropriate functionalities, such as VDI updates in the event of changes in price or product availability. They will also support context-aware applications, such as interoperability with GPS devices in order to guide consumers to a specific sales location.

9.5.2 Users

The main users identified to this scenario are:

- Manufacturing companies who wish to publish information on their products;
- Consumers who want to shop more easily;
- Store owners who wish to publish VDIs for their list of products and thus increase their sales;
- Customer protection organizations.

9.5.3 Products and Services

In this scenario some products and services could be provided, namely in the following areas:

- CoNet certification to manufacturers;
- Product VDI retrieval;
- Interoperability with context-aware applications, such as GPS navigators.

9.5.4 Business Flow

The enterprise which could invest on smart product will be enabled to offer its customers a broad range of services for the integration and customization of retail solutions.

CONVERGENCE will add another service to our portfolio, which already includes offering from several leading technology providers including Oracle. Our main strategy for selling to existing customers will be to show them examples of

CONVERGENCE solutions for the retail business, demonstrating their benefits, primarily in terms of cost reductions.

CONVERGENCE is a disruptive technology, with no direct competitors. In these conditions it is premature to consider pricing for CONVERGENCE-based retail solutions on smart products.

The Fig. 9.5 shows the relevance of certification in the product manufacturing and distribution chain. The certification will be pervasively requested as smart products will be available on the market.

9.5.5 Comparison with Traditional Implementations

Customers do not have an easy way to obtain all relevant information for a product. A common practice is to visit the product's webpage (if available) and get an idea of the product's characteristics. However, it is unusual that the webpage lists all possible sales locations. With the CONVERGENCE solution, each individual storeowner will be able to subscribe to the CONVERGENCE application and add his store as a sales place for a specific product. The VDI is then updated and the consumer can retrieve this information. Another problem is that in the event of a product recall, there is no immediate way to inform consumers. In our

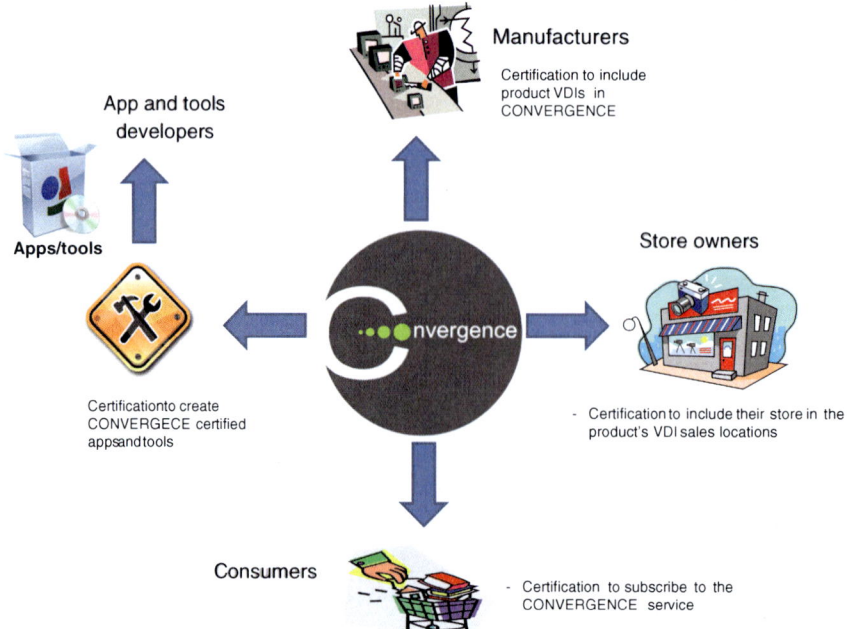

Fig. 9.5 Relationships between CoMid provider and involved members

proposed model, consumer protection organizations would update the corresponding VDI and owners of the product would be instantly informed.

9.5.6 Business Models

The CONVERGENCE software developer sells licenses to manufacturers who wish to publish VDIs concerning their products. Storeowners may also buy a license, so as to be included in the list of sales locations for the product. Individual consumers will pay to use this CONVERGENCE application in their mobile devices because it helps them to shop. Finally providers will seek cooperation with software developers in order to develop the necessary applications.

9.6 Augmented Lecture Podcasts for Education

This business scenario has been contributed by LMU—one of the renowned universities in Germany. As educational institution, LMU aims at educating future generations for research and business. Complementary to traditional teaching, LMU works with contemporary solutions, e.g. podcasts and social learning environments.[2] This has become more and more important in times of technological advances.

9.6.1 Description of Business Scenario

The spread of information and communication technologies have created new possibilities for educational services (at universities), including lecture podcasts. They usually consist of audio/video records that sometimes are accompanied by presentation files. Such podcasts support students during individual learning, allowing them to choose when and at what pace they want to study a lecture. An innovative idea is to augment lecture podcasts within a social learning environment that provides additional features like annotations to enhance student-to-student as well as student-to-teacher communication.

Based on this idea, we present some business ideas for the augmented lecture podcast scenario in which universities offer their services over the CONVERGENCE Network. It shall be noted that universities as educational institutions do not primarily aim at financial gains, but focus on non-financial gains in education as well as research which includes:

---

[2] http://videoonline.edu.lmu.de/

- Attract potential students to enroll to our university;
- Improve the quality of educational services for subscribed students;
- Enhance the university's reputation in terms of international competition as well as research;
- Support of national and international cooperation.

The benefits for these users do not directly come from CONVERGENCE, but rather from the content that is provided through CONVERGENCE. However, those two aspects are tightly coupled with each other. The CONVERGENCE framework offers several built-in functionalities that support these ambitions:

1. CONVERGENCE is a common standard. It enables interoperability and avoids platform lock-in and cooperation;
2. CONVERGENCE maintains the relationship of all copies to the original source. When content is modified, automatic updates are made at the next adequate time. This ensures the quality of learning;
3. CONVERGENCE has an integrated notification mechanism. This enables asynchronous searches. Students do not miss the chance to find relevant content that is not published yet. They are notified as soon as new content is available;
4. CONVERGENCE ensures reliability of service. It allows content caching so that students can be offline while still being able to retrieve content (e.g. from students close to them).

In summary, CONVERGENCE improves the overall service quality of educational services and the learning quality in general. This in turn positively influences the Universities reputation; attracting potential students and allowing us to gain revenues from newly subscribed students.

9.6.2 Users

The main actors of the depicted scenario are students and lecturers. While lecturers function as content producers, students take the role of content consumers as well as producers. They search, download, watch and annotate podcast VDIs.

Currently, the department of Computer Science at LMU has about 1900 students, of which more than 700 are studying media informatics. However, there are also other potential users of CONVERGENCE-based services which include students from other departments and occasional students from other universities and countries who sometimes use the department's learning materials.

9.6.3 Market, Products and Services

There are two potential markets for universities: a local market which comprises students and users within the country, but also a global market, which is bigger in size and which yet has to be developed, but if such global market exists, new sources of revenues will unfold (e.g. from advertising, start-ups, etc.).

Services to Lecturers CONVERGENCE facilitates the production and modification process of podcasts for lecturers—description, licenses and resources are all nested within a VDI. Once this package is published, lecturers do not have to take care about notification mechanisms (this is a built-in feature by CONVERGENCE). The same applies when podcasts are updated.

Services to Students CONVERGENCE improves existing features for students. Students can perform asynchronous searches in order to find content that will be published in the future. They can subscribe to content of their interest and are automatically notified (e.g. when new podcast episodes are released).

9.6.4 Comparison with Traditional Implementations

The augmented lecture podcast scenario does not introduce new functionality to the podcast creation process or to social learning environments, but CONVERGENCE changes how these functionalities are achieved.

There are already a variety of tools that support the podcast creation process. Authoring tools like ProfCast,[3] Camtasia[4] or Lecturnity[5] allow lecturers to produce their own lecture podcast for presentations. A variety of universities run social learning environments for their students. For example, Unterrichtsmitschau[6] at LMU offers an online learning environment where students can watch lecture podcasts and discuss them with other students. MITx[7] is a new initiative to expand the online learning experience. It is supposed to be launched in fall 2012. Last but not least, there is Moodle,[8] an open source content management system for online courses. It can be used for a variety of purposes and many universities use this software complementarily to their actual course.

CONVERGENCE does not replace all these applications, but brings new opportunities for them. The implementations of these applications are quite different from each other, meaning that connecting two applications is cumbersome (platform lock-in) due to the different technologies used. However, having

---

[3] www.profcast.com/

[4] http://www.techsmith.com/camtasia.html

[5] www.lecturnity.de

[6] http://videoonline.edu.lmu.de/

[7] http://mitx.mit.edu/

[8] http://moodle.org/

CONVERGENCE-based applications will provide a common infrastructure for collaboration and content sharing, but still respecting copyrights of content through its rich licensing mechanism. CONVERGENCE changes the way we handle content and how application functionalities are implemented!

9.6.5 Business Models

As an educational institution, LMU's primary mission is not to generate revenue but to attract new students, improve the quality of the educational services the university offers to existing students, and to compete in international research.

We believe that the podcast market is still in movement and will play an important role in the future. The Internet has not only changed the way we consume media, but also the way we learn (look at Wikipedia). Podcasts are part of this trend. Between October and December 2011, Sebastian Thrun (of Stanford and Google) held an online course on Artificial Intelligence that had about 160,000 subscribers from all around the world.[9] Thanks to this success, he decided to leave the university and found a new online learning company, which he has called Udacity.[10]

This anecdote highlights the potential of this kind of podcast. There will be more to come. In the future, it will be difficult for universities to compete with online lectures from top researchers. One strategy could be to incorporate such lectures in their own learning materials, enhancing them in the way they see fit, and encouraging student commentary and discussion on the lectures. CONVERGENCE would be ideally suited to support such a strategy. It has the potential to provide a common infrastructure for this approach to share content between universities, respecting ownership rights with its rich licensing language.

From a commercial point of view, there are further opportunities for augmented podcasts, especially for providers of online courses. Since CONVERGENCE allows continuous updates, providers of augmented podcasts for online courses can extend and renew their business models to include subscriptions. Instead of making one-time payments to download podcasts, customers could pay a subscription and keep their podcast up to date for one, three, 6 months or more. Another possibility would be to offer partial subscriptions, in which a customer subscribed only to particular chapters in a podcast, or only to video or slide content. Of course, these models could be combined. Augmenting podcasts with annotations increases product quality as well as user satisfaction. Customers can easily ask questions, discuss with other customers, and experience a new form of

---

[9] http://blogs.reuters.com/felix-salmon/2012/01/23/udacity-and-the-future-of-online-universities/

[10] http://www.hackeducation.com/2012/01/23/stanford-ai-professor-thrun-leaves-university-to-start-udacity-an-online-learning-startup/

learning in a learning community. Using CONVERGENCE, personnel who administrate discussions can easily remove inappropriate content or support customers with answers to questions.

9.7 Retail In-Store Logistics and Post-sales Services

As one of the largest Romanian integrators and providers of complex information management solutions, UTI is interested in extending its integrated information management expertise to the management of real world objects. UTI Retail already has an extended base of installed security systems to monitor, track and perform surveillance on real-world items. However many of our major customers are starting to demand more sophisticated services. Thus, UTI has started to work on the development of new intelligent real-world objects management services that offer enhanced benefits to different types of actor (retailers, customers, brand owners, etc.). The CONVERGENCE project complements the company's efforts in this direction.

9.7.1 Description of Business Scenario

This business scenario focuses on the use of VDIs in the retail business, where they can be used both for in-store logistics and post-sales services. Any retailer with a RFID/barcode infrastructure can associate (selected) products with a VDI and use CONVERGENCE to build extended software services to manage the product lifecycle inside the store. For example, it would be possible to create services that monitor warehouse stock and automatically place a purchase order when the stock falls below a certain threshold or when it is falling very rapidly. CONVERGENCE can also provide product placement services that offer information on the status of VDIs on the shelf (number of current items, last check-outs, last check-ins etc.) and trigger alarms if the stock on the self falls below a threshold. Customers can benefit from a better shopping experience at the shelf (access to extra product information).

The second category of retail business services that can be implemented using the CONVERGENCE infrastructure is post-sales services. For example, the retailer can save time and money by automatic issuing digital warranties and other product related documents in the form of VDIs. Customers can use the product VDI to search for associated services (accessories, upgrade, repair) and for the best providers of these services, possibly saving money. Finally, different service providers can use the CONVERGENCE platform to advertise their services through VDIs and increase their sales and businesses.

9.7.2 Users

The main users in this business scenario are:

- Retailers: use the VDIs for better in-store and post-sales services;
- Customers: use the VDIs associated with products they (wish to) buy to find additional information and services;
- Service providers: use the CONVERGENCE platform to advertise their services.

9.7.3 Products and Services

UTI developed two types of CONVERGENCE-based services:

- Services for consumers:
  - Searching for products and subscribing to products of interest using as many description attributes as possible (if desired)
  - Receiving a notification when a retailer publishes an offer which matches the customer's preferences
  - Receiving in-store notifications of personalized offers and local/in-store deals.
- For retailers:
  - Using VDIs to publish offers that reach customers outside the retailer's own ecosystem
  - Monitoring the interests expressed by consumers
  - Monitoring matches between customer preferences and the retailer's offers
  - Monitoring consumer behaviour inside and outside the store.

9.7.4 Table of Transactions

The Table 9.3 below presented the predictable financial rewards of this scenario.

Table 9.3 Cash flow and financial indicators for UTI smart retailing service

	Y1	Y2	Y3	Y4	Y5
Revenues	46,770,00	102,750,00	142,750,00	200,920,00	263,900,00
Costs	147,200,00	127,650,00	100,050,00	100,050,00	100,050,00
Revenues—costs	−100,430,00	−24,900,00	42,700,00	100,870,00	163,850,00
Accumulated cash flow	−100,430,00	−125,330,00	−82,630,00	18,240,00	182,090,00
Internal rate of return—33,082 %					
Net Present Value at 10 %—90,836,16 EUR					
Break-even—Y3					

9.7.5 Business Flow

UTI, as a technology integrator provider will play an important role in facilitating the actors in the retail chain access to the CONVERGENCE technology. More specifically, UTI will help retailers to enable their technology to CONVERGENCE by:

- Offering development services to retailers to implement the connectors/adaptors to the CONVERGENCE platform;
- Help retailers advertise product offers in the VDI format;
- Help retailers subscribe to manufactures feeds (expressed as VDIs) about new product releases, product updates and other news;
- Adapt their technologies to connect to other service provides that manage product VDIs (Fig. 9.6).

9.7.6 Comparison with Traditional Implementations

From the retailer point of view, RFID/barcode systems that provide product management services are already in widespread use. However, these systems are generally custom developed and are not interoperable. Thus, the retailer is unable to offer a standard interface to its business partners (manufacturers, providers, customers, other service providers.). CONVERGENCE, by contrast will allow retailers to exchange information with all these actors in a standardized way.

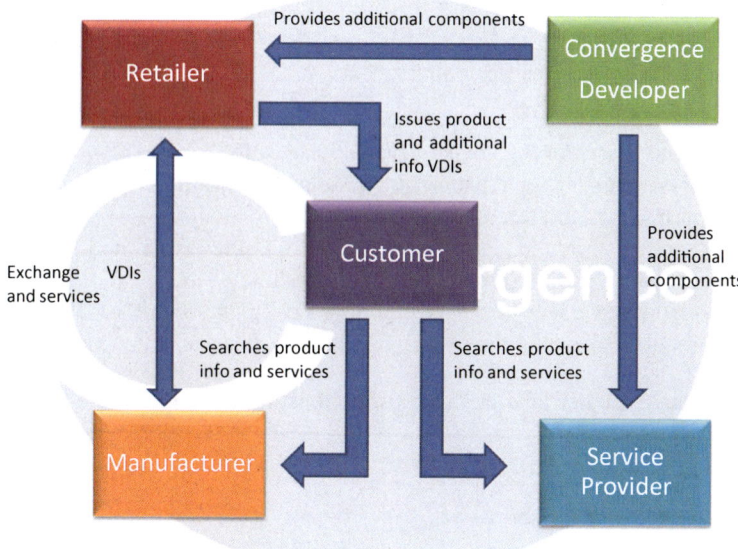

Fig. 9.6 Business flow in UTI business scenario

The retailer will be able to easily publish, search and subscribe to relevant information, thus saving time and money through interoperability.

The same advantages apply to customers. In today's world, a customer doesn't have a standard place to look for information and services related to a product she buys. All she can do is browse the Internet and use search engines to find product details, manuals, comments or repair shops. By contrast, VDIs associated with a product, can provide her all the information she needs. Finally, there is no market today where different service providers can advertise their services and compete to offer the best alternative to their clients. CONVERGENCE will create such a market, allowing them to access a potentially large customer base.

9.7.7 Business Models

UTI is considering several different business models.

9.7.7.1 Platform as a Service-in-the-Cloud

One model involves running the platform as a hosted service using a range of different charging mechanisms, as specified below.

Monthly Subscriptions

Monthly Subscriptions services mainly target brick-and-mortar stores. UTI envisages four types of subscriptions:

- Free—up to 5 advertisements per month, unlimited email notifications based on matches, no discriminated ranking in search results;
- Standard—up to 50 advertisements per month, unlimited email notifications based on matches, preferential ranking in search results;
- Silver—up to 200 advertisements per month, unlimited email notifications based on matches, preferential ranking in search results;
- Gold—unlimited advertisements, unlimited email notifications based on matches, preferential ranking in search results.

To facilitate retailer take-up, all subscriptions in the first 6 months will be free.

Pay Per Click

PPC services will target online retailers. PPC display advertisements will be shown on a Smart Retailing web site with related content and will receive preferential ranking in search results. Retailers will pay when customers click on the link present in a list of search results.

To facilitate retailer take-up, services in the first year will be free.

Sponsored Links/Banner Ads

In addition to providing search results, the Smart Retailing web site will also include sponsored links. These will be clearly identified and kept separate from search results. Some will be context-based (based on customer profiles); some will not. Retailers will pay by "impression" (each time a customer views an ad). The site will never display more than six sponsored links/banner will be displayed simultaneously.

In-store Digital Campaigns

The platform will be designed for deployment to different types of terminal including mobile terminals. It will thus be possible to use it as the basis for in-store digital campaigns.

9.7.7.2 Platform as Integrated Solution

UTI is also considering the possibility of licensing customized versions of the platform to individual retailers. In this case, revenues would come from:

- Licensing fees;
- Customization;
- Maintenance and support.

UTI sees the first business model primarily as a short-medium term solution. In the longer term, we expect ICN services to become a standard part of the Internet. Once this occurs our dedicated CONVERGENCE cloud would lose its market.

In the longer term, UTI sees itself as a company that earns revenues by helping companies to exploit future Internet technologies (including CONVERGENCE or other ICN-based technologies).

9.8 Services for a Product-Supply-Chain

This section describes the market opportunity for retailers who require efficient match of available products and consumer needs. In particular Wipro is willing to exploit by integrating between CONVERGENCE and Oracle Retail solutions, thus enhancing the services WIPRO provides to retailers. Wipro consultants and technicians have a proven track of expertise and success in different retail areas such as customer analytics, RFID, global data synchronization, in-store merchandizing and pricing, and supply chain management with focused solutions for the grocery, pharma and specialty retailing segments.

WIPRO will use CONVERGENCE to provide new services to retailers, earning additional revenues by supporting these services. The services WIPRO will provide offer customers a number of important innovations.

- A standard digital container (VDI) for product information, digitally signed by the manufacturer;
- Retailer and consumer access to up-to-date product information distributed via the CONVERGENCE network;
- Integration between CONVERGENCE technology and existing retail tools and frameworks;
- Facilitated migration of product information from VDIs to retailer systems and databases;
- Improvements in in-store product management (product recalls, warranty verification, control, fraud and piracy prevention, etc.);
- CONVERGENCE-enabled POS machines, providing digital receipts and warranties;
- CONVERGENCE-enabled retail applications that operate on top of CONVERGENCE middleware.

9.8.1 Description of Business Scenario

The CONVERGENCE Project can only achieve its full potential if it offers the tools necessary to generate revenues. These are the laws of today's world: a great project comes from a great idea, but its development is directly proportional to the money it generates. Against this background, the business scenario described here focuses on a single issue. How can CONVERGENCE help retailers make money? If we want retailers to support CONVERGENCE, this is the question we have to answer.

In this business scenario we consider a VDI-enabled product supply chain, where a VDI is associated with a product after it has been manufactured and the information provided on the VDI is certified by a security entity that is paid to perform this service. In the scenario, product manufacturers use the VDI for cross-merchandising and cross-marketing. To do this, they enhance the VDI for a product with information about other similar or compatible products. This can become a huge marketing tool allowing them to make their products known to consumers through the VDIs they create or own. Cross marketing could involve multiple brands. For example, an iPod VDI could have information about compatible speakers from Logitech or Sony; insurance companies and repairers could pay to have their service information available in the VDIs for selected products. Finally, a VDI-enabled product supply chain could also include second-hand sales. The CONVERGENCE Platform could be a splendid substitute for websites like eBay. In this scenario, all a consumer would have to do is publish the VDI for the product on the platform. This would represent a big step forward in terms of reliability and trust.

9.8.2 Users

The main considered users for this scenario are:

- Manufacturers: Product manufacturers that create VDIs, using CONVER-GENCE tools. Manufacturers are also responsible for associating their products VDIs with other products or services;
- Retailers: Retailers can also associate VDIs with their other products and services (e.g. VDIs for, insurance companies or repairers they trust;
- Consumers: From the point of view of the supply chain consumers are end-users. However they can also become sellers, advertising VDI-enabled products they own on the CONVERGENCE Second-Hand Sales Platform;
- Security Entity: Companies responsible for certifying the information contained in VDIs, increasing consumer trust;
- Insurance Companies/Repairers: Entities that try to make deals with manufacturers and retailers so they can associate their services with product VDIs.

9.8.3 Products and Services

Some pertinent products and services can be considered for this scenario:

- VDI creator: A tool allowing users to create and update VDIs;
- CONVERGENCE Certification Service: A tool allowing security entities to certify the VDIs information;
- CONVERGENCE Search Engine: A tool allowing anyone to browse VDIs according to search criteria and user restrictions;
- CONVERGENCE Sales Service: A tool allowing anyone to publish their VDIs for second-hand sale, making them available in the CONVERGENCE Second Hand Sales Platform.

9.8.4 Business Flow

Wipro Portugal offers its customers a broad range of services for the integration and customization of retail solutions.

CONVERGENCE will add another service to our portfolio, which already includes offering from several leading technology providers including Oracle. Our main strategy for selling to existing customers will be to show them examples of CONVERGENCE solutions for the retail business, demonstrating their benefits, primarily in terms of cost reductions. However, we expect it will be easier to sell to new customers, who have no investment in previous WIPRO solutions.

One of the biggest obstacles we will have to face is that WIPRO focuses on retailing and does not work directly with manufacturers or with other stages in the supply chain.This means other actors will have to cover the manufacturer segment of the supply chain.In brief, the plans of WIPRO to commercialize CONVER-GENCE-based solutions require a pre-existing CONVERGENCE infrastructure. WIPRO recognizes this as a weakness in its plans.

CONVERGENCE is a disruptive technology, with no direct competitors. In these conditions it is premature to consider pricing for CONVERGENCE-based retail solutions. If WIPRO decides to offer such solutions to its customers, the prices will be confidential, and adapted to the needs of the customer, as is the case for all WIPRO products and services (Fig. 9.7).

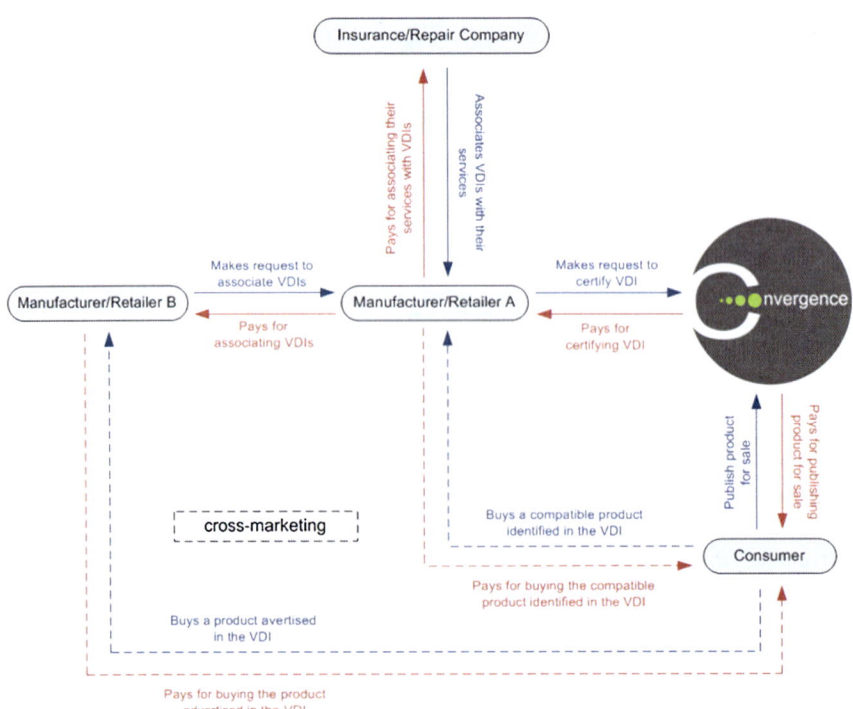

Fig. 9.7 Relationships and revenue flows between a provider of CONVERGENCE services and involved members

9.8.5 *Comparison with Traditional Implementations*

The business scenario described here offer a massive improvement compared to current implementations. Nowadays we are facing a big problem in terms of information reliability, every day we are confronted with large quantities of information, but we don't know for sure which information we should trust. With VDI signed by its issuer people will know the source of. This is especially important for the second-hand sales market, where users will know in advance that they can have trust the description of the product they are buying. Instead of a product description they will have access to a certified VDI containing all the product's data.

In cross-merchandising and cross-marketing, CONVERGENCE will provide a great tool for products manufacturers and retailers. The marketing potential of associating product's VDIs is incalculable when compared to the tools available nowadays. For consumers, VDIs can resolve many problems like not knowing which speakers are compatible with an iPod, which are the best routers for a modem or what graphics card to buy for a laptop.

CONVERGENCE will also make a big contribution to the market for product insurance and repair, helping not just insurance and repair companies but also consumers. With CONVERGENCE, it will be much easier for a consumer to access the available options to obtain insurance for her product or to repair if it is broken. It will also be much simpler to get products repaired CONVERGENCE will give us confident consumers.

9.8.6 *Business Models*

The main business models proposed by Wipro Portugal are:

- Certifying VDI information: The CONVERGENCE Platform will provide a security infrastructure that will operate as a broker between CONVERGENCE and product manufacturers. When creating a new VDI for a specific product, manufacturers will use the CONVERGENCE platform to request a service provider to sign the VDI. The service provider will then provide a special signature for the VDI if the information in the VDI is trustworthy;
- Associating VDIs with other VDIs (cross-merchandising and cross-marketing): The CONVERGENCE Platform will provide a service that allows manufacturers and retailers to incorporate links to other VDIs within the VDIs for their own products and thus to support cross-merchandising and cross-marketing DIs;
- Associating VDIs with insurance and repair companies: The CONVERGENCE Platform will provide a service that permits manufacturers and retailers to associate their VDIs for their products with services offered by insurance and repair companies. When the companies have reached an agreement, they can use

the CONVERGENCE Platform to include the necessary links in the VDIs for
their products;

- Second-hand sale of VDI-enabled products: The CONVERGENCE Platform
  will provide an option making it easier for consumers to sell their VDI-enabled
  products second hand. When a consumer owns a VDI-enabled product and
  wants to sell it, all she has to do is access the CONVERGENCE Platform and
  publish its product VDI. Other users will then be able to find the product via the
  CONVERGENCE Search Engine.

9.9 Marketing and Advertising in Publish/Subscribe Networks

CONVERGENCE can be a powerful instrument in the marketing and advertising
activities of an enterprise. Enterprises can be enabled to distribute their applications, and nowadays applications are used as entry point to real shops or e-commerce (see IKEA for furniture, Amazone.com for books and many other product
categories, etc.). As more applications can be distributed offering specialized
services, more often the marketing and advertising units use them to present new
products or support the choice of the clients.

9.9.1 Description of Business Scenario

The general rationale for this scenario is based on a preliminary analysis of the
impact of the publish/subscribe paradigm on marketing as compared to current
marketing models.

Today, actors who want to publicize their products need to establish business-
to-business relationships with other actors (e.g. newspapers, magazines, broadcast
media, internet search engines, email spammers) who can carry their
advertisements.

CONVERGENCE will not replace this model. However it has the potential to
create new channels of communication between producers and consumers and new
business models.

9.9.1.1 Commercial Advertisement Through VDIs

The CONVERGENCE framework allows advertisers to inject advertisements into
the network in the form of VDIs.

In this scenario, we take the YouTube marketing model, in which Mercedes Benz posts a video commercial for its latest "Blue Engine" to YouTube with the tags CARS, ECOLOGY, and use CONVERGENCE to apply it to the network as a whole. Unlike the traditional model, CONVERGENCE makes it possible to give a precise classification to the advertisement, making it easier for users to find what they are looking for. It also allows advertisers to clearly indicate when a particular item of content should be classified as "commercial information"—a legal requirement in many countries.

Skilled marketing departments will use CONVERGENCE to create commercials that comply with the law while exploiting the enhanced possibilities offered by the new system.

9.9.1.2 Trading Information about Users Interests/Subscriptions

The CONVERGENCE subscription model allows users to express their habits, preferences, tastes, and requests in a precise machine-readable format. However, making this possible at a technical level, does not guarantee that most people will do it. User-friendly design patterns at the Application level (such as the photo tagging tool of Facebook) must be implemented in order to facilitate the widest acquisition of context information. For marketing businesses, the knowledge that the user has expressed a clear interest in a specific topic is worth gold and is in fact much more valuable than the information they can extract from search requests to Internet search engines such as Google. It is easy to envisage an ecosystem of technology companies specialized in building software spiders to crawl CONVERGENCE nodes, collect subscribe requests, aggregate them and sell the data to marketing businesses or marketing departments.

We are fully aware that this scenario raises major issues of user privacy. To better shed light on its several implications, the following observations are to be underlined:

1. We do not embed any private or personal information about users inside the subscriptions they perform, that may allow to link back to one specific identity. When results are available and match a subscription, they are routed back to the CONVERGENCE peer through which the subscription was issued, not to a specific user. Only the Notification Service of that peer knows how to reach the user. Thus, subscribe information that circulates inside the system is decoupled from info about who formulates subscriptions;

2. The approach of CONVERGENCE is not to devise technical solutions able to ban these businesses from going on. Since CONVERCENCE acts as a powerful search and retrieve engine, it is obvious that this knowledge will be exploited commercially, also. We seek to distribute this possibility to all people using it, since the system is distributed. Aggressive data mining strategies are hence more difficult to be employed, compared to concentrators such as Google or Facebook. CONVERGENCE is designed to be transparent, have a public API

to hook to its knowledge, and to manipulate that data according to existing (and novel) privacy legislation;

3. The VDI, which is the basic self-contained information block circulating among CONVERGENCE peers, follows an open standard that defines its structure and format. CONVERGENCE keeps this standard and its source code open and compliant to the law, allowing refining its privacy preserving capabilities from the feedback of the received attacks or public discussion about technical choices;

4. CONVERGENCE devises a flexible technical solution. If a provider wants to offer a closed service or application on top of CONVERGENCE middleware, hence asking its users to fully identify themselves on the system (e.g. a ticket selling infrastructure where personal information is needed in order for the service to be carried out, and focused advertisement about Opera Houses is conveyed to users who gave consent), then it might as well encrypt everything and decide to deploy specific gateways to be interoperable with the rest of CONVERGENCE. Inside its network, personal information about users tastes will be non searchable by external spiders.

Furthermore, this scenario offers the possibility to test CONVERGENCE in a potentially controversial use case, where the capabilities of the system and its precision can be fully evaluated on the background of their commercial value.

9.9.1.3 Related Business Scenarios

CONVERGENCE's own search and subscription solutions are not necessarily the best solutions imaginable. Once the system APIs of the CONVERGENCE middleware are clearly defined, technology providers could create their own independent search engines for CONVERGENCE, perhaps with different ranking and grouping algorithms.

9.9.2 Users

The main users of this scenario are:

- Marketing departments of companies;
- Developers of data mining solutions.

9.9.3 Products and Services

Some relevant products and services were also considered:

- VDIs containing commercials or linking to them;

- Data about subscriptions: marketing companies extract and sell aggregated data about users' interests;
- Subscriptions Crawler: a spider that enters CONVERGENCE, and exploits data aggregation algorithms and data mining algorithms while scanning the flows of subscriptions entering the system.

9.9.4 Business Flow

The Fig. 9.8 represents the business flow where the marketing department of an enterprise can purchase the license for advertisement VDI or accessing data about user's subscriptions while technology providers can acquire aggregated data about user's interest.

9.9.5 Comparison with Traditional Implementations

Like today's web, CONVERGENCE can be used as a worldwide advertising platform. However, CONVERGENCE offers several new benefits: it makes it easy to target advertisements to users with specific interests, when they give consent, and easy for users to find them. It also makes it easy to filter out advertisements where this is required by law. With CONVERGENCE users will only receive information on their explicit request. This will limit spam and contribute to better and focused advertising.

If the depicted business scenario is to work on a healthy basis, then several issues have to be tackled in the CONVERGENCE architecture, such as:

- In order to protect the user, CONVERGENCE Applications that allow Subscribe actions should be carefully designed by developers, in order to prevent malware or bots subscribing to unwanted contents on behalf of the user;
- Content caching on multiple nodes should not prevent Law Authorities from removing or obfuscating unwanted advertisements;

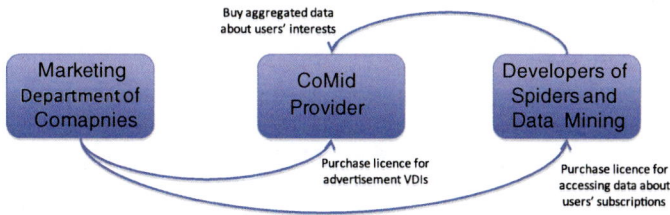

Fig. 9.8 Transactions in the "Marketing and advertising in publish/subscribe networks" business scenario

- Spiders crawling the CONVERGENCE node and collecting precious data from the flow of subscriptions should be regulated and should only be able to access information through a proper API;
- Private information shall always be aggregated and unrelated to specific users.

9.9.6 Business Models

Business Models for this scenario focus on trading and buying information from information providers that collect it from CONVERGENCE nodes. The information traded in this way is potentially of higher value than the information collected by other concentrators such as Google. CONVERGENCE can thus change the shape of the market. At the same time, the new possibilities are likely to require new regulations to protect user privacy.

The use of VDIs to carry advertisements offers a new potential source of revenue for CoMid providers. Given the need to meet special legal requirements and technical specifications, providers could charge more for advertisements than for regular VDIs. Given that VDIs will have be better targeted and have higher user acceptance than other forms of advertising marketing businesses will be able to charge higher fees than for traditional advertising. Finally the availability of information on users (subject to regulations and/or business-to-business relationships) can provides another source of significant added value.

In CONVERGENCE, very specific licensing and business-to-business models can be developed, that will depend on the size of the CONVERGENCE network and the size of potential target audiences.

9.10 Licensing Schemes for the CONVERGENCE Software

There are several licensing options CONVERGENCE project could address for the technologies it developed. The BSD license is the least restrictive scheme: The license allows the use of BSD-released source code even in projects that do not use the BSD license, including commercial closed source software. Being a permissive, free software license, it places minimal restrictions on the redistribution of the software. This is in contrast to copyleft licenses, such as the GPL family which have reciprocity/share-alike requirements.

The Mozilla MPL license strives to stay in the middle: it allows covered source code to be mixed with other files under a different, even proprietary license. However, code files licensed under the MPL must remain under the MPL and remain freely available in source form.

All pre-CONVERGENCE MPEG-M middleware components were released under the BSD policy. Thus, the main decisions the partners have to take concern the new components the project has developed. These include:

- New MPEG-M middleware engines (Overlay, Match, CoNet, protocols, …);
- Tools;
- Applications;
- The network-level CoNet core.

The partners believe that MPEG-M made the right choice with BSD, since the focus is on protecting the specification, and not the code. More generally, the project agrees that CONVERGENCE should protect its intellectual property by means of scientific publications, where appropriate, and freely release the derived source code.

The following Table 9.4 summarizes the licenses the partners will use to protect their software and the rationale for its choices.

The project wishes to stress that the CONVERGENCE Tools represent a major contribution to the MPEG-M community, facilitating integration between Applications and the MPEG-M eco-system of engines. By releasing them under BSD licensing (vs. LGPL, for instance) CONVERGENCE guarantees that the MPEG-M community can share all the benefits. This licensing solution avoids the risk that CONVERGENCE Tools (and Applications running on top of the tools) could be isolated from the engines, and promotes the introduction of novel Tools.

Table 9.4 Licensing choices for CONVERGENCE products

	License type	Motivation
New MPEG-M engines	BSD	Stay in line with MPEG-M
CoMid tools	BSD	The CoMid tools make it easier to use the MPEG-M middleware and address the same user community. It is appropriate, therefore, that they should be released under the same licensing scheme
CoNet core	LGPL + GPL	This choice is dictated by the licensing of the original CCNx project. CONVERGENCE modifications to the network forwarding functionalities of CCNx will therefore remain under GPL. New functionality for the routing plane will be licensed under LGPL (allowing middleware wrappers to remain under BSD)
CONVERGENCE Peer Kit (CPK)	Mozilla + BSD	This bundle includes a reduced set of CoMid engines and a simple demo Application
Applications	Will be decided by the owners of individual applications	This allows enterprises to protect their investment when building CONVERGENCE-compliant software

There are no issues with the CoNet TE (an MPEG-M engine) being licensed under BSD, although it uses our modified version of CCNx. CCNx is partially licensed with a GPL strict scheme (core networking stack running on routers) and partially with LGPL (API library). The creators of the CCNx package have used this scheme on purpose. Coherently, CONVERGENCE adheres to the same mixed scheme for its own newly developed CoNET routing functionality, extending and enhancing CCNx.

As far as concern Applications, owners of CONVERGENCE-compliant applications can apply whichever licensing scheme they choose to suit their needs. The link to the middleware does not impose any special restrictions. For instance, trial applications developed by UTI and WIPRO will not be completely open sourced, and will remain the property of the developers; other developers involved in the CONVERGENCE trials have decided to release the source code of their applications to the community.

For instance, the demo Application contained within the CPK is released under a MOZILLA public license, but the CPK uses the MPEG-M middleware, which is released under a BSD license.

Except where prohibited by licenses for third party software, incorporated in CONVERGENCE, CONVERGENCE will adopt the MPEG BSD license. The consortium has made available part of the sources developed during the project through the CONVERGENCE web site (http://www.ict-convergence.eu/demo downloads/).

9.11 Conclusions

Keeping converged networks, services and devices up and running perfectly, while being able to capture and bill for them, will be a major differentiator among operators and players. In that sense, broadband, IP adoption and converged services dissolves traditional barriers by enabling new value chains to take shape. At the same time, for those industry players who get it right the same physical and software infrastructure will bind them together in ways never seen before.

Researchers around the world are studying how this change will affect the society and businesses. There are relevant advantages from the way contact centre agents work and interact with customers, to the way we work from home or on the move as effectively as if we were at a desk, to the way that businesses become more efficient and integrated communicators.

What looks clear is that we are at the beginning of a huge change in the way business is done. As we demonstrated in this chapter, CONVERGNECE services and tools can be exploited in different scenarios that will improve employee productivity and customer or citizen satisfaction.

The academic partners will continue to participate in standardization activities and to support the CONVERGENCE framework and the commercial partners will continue with their individual exploitation plans.

Chapter 10
Conclusions and Future Research Topics

Nicola Blefari Melazzi, Teresa Andrade, Richard Walker, Leonardo Chiariglione, Iakovos S. Venieris and Heinrich Hussmann

Abstract The final outcome of the CONVERGENCE project is the full specification and an Open Source implementation of the CONVERGENCE system, which comprises the CONVERGENCE network and middleware levels, as well as a set of prototype applications demonstrating the functionality of the network and middleware in different business scenarios.

The final outcome of the CONVERGENCE project is the full specification and an Open Source implementation of the CONVERGENCE system, which comprises the CONVERGENCE network and middleware levels, as well as a set of prototype

N. Blefari Melazzi (✉)
Dipartimento di Ingegneria Elettronica, Università degli Studi di Roma Tor Vergata, Roma, RM, Italy
e-mail: blefari@uniroma2.it

T. Andrade
Telecom and Multimedia Division, INESC Porto Campus da FEUP, Porto, Portugal
e-mail: mandrade@inescporto.pt

R. Walker
Blue Brain Project, Ecole Polytechnique Federale de Lausanne, Lausanne, Switzerland
e-mail: rwalker@xiwrite.com

I. S. Venieris
Intelligent Communications & Broadband Networks Laboratory, ICBNet,
School of Electrical & Computer Engineering, 9 Heroon Polytechniou str,
Zografou Campus, Athens, Greece
e-mail: venieris@cs.ntua.gr
URL: www.icbnet.ntua.gr

H. Hussmann
Institut für Informatik, Ludwig-Maximilians-Universität München, München, Germany
e-mail: Heinrich.Hussmann@lmu.de

L. Chiariglione
CEDEO.net, Via Borgionera, 103 I-10040 München, Italy
e-mail: leonardo@chiariglione.org

F. Almeida et al. (eds.), *Enhancing the Internet with the CONVERGENCE System*, 263
Signals and Communication Technology, DOI: 10.1007/978-1-4471-5373-3_10,
© Springer-Verlag London 2014

applications demonstrating the functionality of the network and middleware in different business scenarios. Additionally, it includes the results from trials, simulations and models demonstrating the scalability, robustness and security of the overall system, its applicability to real-life business scenarios and its acceptability to users.

The results delivered by the project are sufficient for third party developers to design and deploy their own CONVERGENCE-compliant applications, supporting dedicated high-level requirements different from those identified in the project, but taking full profit of the underlying content-based, publish-subscribe functionality of CONVERGENCE. In addition, several of the results of the project have been contributed to standardization and some of them are now International Standards.

From an end-user point of view, CONVERGENCE offers new ways of handling digital resources. Focus groups suggested that the most important "selling points" are transparent synchronization of content among multiple users, fine-grained author control over access rights and payment conditions, the grouping of multiple information resources in a single self-contained package, built-in support for security and privacy, including "digital forgetting", "democratic" CDN (i.e. CDN functionality available for all) and improved performance. More in general, the project showed that all the promised advantages off the ICN paradigm (see the preface of this book) can indeed be reached.

The CONVERGENCE partners believe that these concepts and their technical implementation in CONVERGENCE will have an important influence on the future development of the Internet and online applications.

In terms of technology, the project contributed to answering important questions concerning the best way to overcome the shortcomings of the current Internet architecture—in particular the mismatch between an architecture focused on communication between hosts and the needs of applications, increasingly focused on content. CONVERGENCE has also tested strategies for dealing with these problems, comparing the advantages of an incremental approach to more radical strategies and opting for the former.

The trials have tested the technical functionality provided by CONVERGENCE in four user scenarios; two have tested CONVERGENCE as a support for a large photo archive and a large archive of video material respectively; a third has tested CONVERGENCE in an educational setting; the fourth and last scenario has shown how CONVERGENCE concepts and technologies can be applied in "smart retailing". In addition, the project decided to implement other two applications, by integrating the four main applications; the first integrated application merges the first (video) and fourth (podcasts) original applications; the second integrated application merges the second (pictures) and third (retail) original applications. The aim of the integrated applications is to show that our system is flexible enough to combine different applications in one and to exploit common VDIs. In this context, we believe that the interoperability demonstrated in the integrated application framework is an important result of the project. The same VDIs are indeed shared and used by diverse applications.

We can conclude that our project has been among the pioneers of ICN; we have been instrumental in defining and implementing a working ICN platform, we published widely on the subject and made several important contributions to standardization and we are well known in the international community, having been invited to numerous events as speakers, keynote speakers and presenters of demos and test-beds and to participate in technical program committees and guest-editorial board of related publications.

All in all, we have been true to the project's main goal "Enhancing the Internet with an information-centric, publish-subscribe service model, based on a common container for any kind of digital data, including representations of people and Real World Objects".

Notwithstanding all this work and the work of other research projects belonging to a first wave of endeavours on ICN, and notwithstanding several papers and specialized workshops that are now at their second or third edition (e.g. SIG-COMM's Information-Centric Networking workshop and INFOCOM's NOMEN), research and development activities on this topic are far from being concluded. Standardization efforts on ICN networks are at their beginning as well, as evidenced by the creation of an IRTF working group (Information-Centric Networking Research Group—ICNRG), which started to draft requirements and scenarios documents on ICN.

Open research topics regarding ICN at large can be classified in four broad areas: (i) functionality of the network/transport layer, the very core of ICN; (ii) functionality of a middleware layer; we believe that some functions and services like content-based publish/subscribe services or semantic search or advanced security/privacy/DRM management are too complex to be implemented at the network level, at line speed, in every network node; we believe that trying to do so would force designers to choose rather simple solutions; thus, we believe that such functions are better performed at the middleware layer, in a subset of all network devices; (iii) applications, related APIs, scenarios and use cases facilitated by the ICN paradigm, e.g. video delivery, Internet of Things, fragmented networks, disaster scenarios; (iv) transversal and general issues, such as: migration path from the current TCP/IP paradigm, allowing also future evolvability; metrics and tools that make it possible to evaluate ICN implementations; business, legal and regulatory frameworks; relationships with supporting technologies, like Software Defined Networking and Network Functions Virtualization; energy efficiency at all layers, including applications.

The first area, functionality of the network/transport layer, which is very important for our field of investigation, can be broken down in more specific research topics, which correspond also to fundamental components of an ICN infrastructure. People and projects working on ICN made a very good start, but are still far to put the word end to research on the following issues:

(1) Primitives and interfaces, which define the relationship of the ICN protocols with the overall architecture.

(2) The naming scheme, which specifies the identifiers for the information addressed by the ICN. The choice of a naming-scheme impacts different aspects an ICN, including the handling of name uniqueness and trademarking, routing scalability, flexibility of supporting different applications, usability, security.

(3) *Name resolution mechanism* a node routes by-name a request of an information item towards a selected serving device. Hence, the ICN node should resolve the name of requested items in the "physical" address of next ICN node towards the selected serving device. Several name-resolution approaches are possible, ranging from an off-path resolution, e.g. based on DHT, to an en-route hop-by-hop resolution exploiting name-based routing tables.

(4) *Routing scalability* with respect to IP, the routing plane of an ICN has to handle a number of information items and corresponding names that is much bigger than the number of IP network prefixes. This has implications on the size of ICN routing tables, on the complexity of lookup functions and on the distribution of ICN routing information and is one of the main concerns of ICN.

(5) *Information delivery* an ICN addresses information items rather than hosts, so it needs a technique to route back the requested information item from the serving device to the requesting device. Some architectures propose to use plain IP means, other ones propose to face the issue through ICN's own means.

(6) *Segmentation and transport mechanisms needed to* (i) split a whole content (e.g. a VDI) in different data units (or chunks); each chunk is an autonomous data unit with embedded security and addressable by the routing plane; (ii) ensure a reliable transfer of chunks from the origin node (or from a cache node) towards the requesting node; (iii) counteract congestion.

(7) *In-network caching* ICN nodes may cache chunks of information. Differently from traditional HTTP caching, an ICN is a *cache network*; this implies the need of properly devising a replication strategy that optimizes the caching space, e.g. by avoiding excessive duplication of content in the network caches.

(8) Resilience against fragmentation of the network.

(9) Energy efficient operation.

(10) *Security and privacy issues tackling (at least) three specific aspects* (i) how to guarantee content authenticity and protect the network from fake content, which could also pollute network caches; (ii) how to guarantee that content be accessed only by intended end users, and (iii) how to protect information consumers from profiling or censorship of their requests.

The interest of the international community on ICN has been always increasing in the last months and we are confident that this exciting field will provide significant theoretic and technologic advances and real world developments and applications, in the near future.

Glossary

Access Rights Criteria defining who can access a VDI or its components under what conditions.

Advertise Procedure used by a CoNet user to make a resource accessible to other CoNet users.

Application Software, designed for a specific purpose that exploits the capabilities of the CONVERGENCE System.

Business Scenario A scenario describing a way in which the CONVERGENCE System may be used by specific users in a specific context or, more narrowly, a scenario describing the products and services bought and sold, the actors concerned and, possibly, the associated flows of revenue in such a context.

CA Central Authority.

CCN Content Centric Network.

Cl_Auth_SC Client Authentication with Smart Card (Challenge Response).

Cl_Auth_User_Pw Client Authentication with Username and Password.

Clean-slate architecture The CONVERGENCE implementation of the Network Level, totally replacing existing IP functionality. See "Integration Architecture", "Overlay Architecture" and "Parallel Architecture".

CoApp The CONVERGENCE Application Level.

CoApp Provider A user providing Applications running on the CONVERGENCE Middleware Level (CoMid).

CoMid The CONVERGENCE Middleware Level.

CoMid Provider A user providing access to a single or an aggregation of CoMid services.

CoMid Resource A virtual or physical object or service referenced by a VDI, e.g. media, Real World Objects, persons, internet services. It has the same meaning of "Resource" and it is used only to better specify the term "Resource" when there is a risk of a misunderstanding with the term "CoNet Resource".

F. Almeida et al. (eds.), *Enhancing the Internet with the CONVERGENCE System*, Signals and Communication Technology, DOI: 10.1007/978-1-4471-5373-3, © Springer-Verlag London 2014

Community Dictionary Service (CDS) A CoMid Technology Engine that provides all the matching concepts in a user's subscription, search request and publication.

CoNet Provider A user providing access to CoNet services, i.e. the equivalent of an Internet Service Provider.

CoNet Resource A resource of the CoNet that can be identified by means of a name; resources may be either Named-data or a Named service access point.

Content-based resource discovery A user request for resources, either through a subscription or a search request to the CONVERGENCE system (from literature). See "subscription" and "search".

Content-based Subscription A subscription based on a specification of user's preferences or interests, (rather than a specific event or topic). The subscription is based on the actual content, which is not classified according to some predefined external criterion (e.g., topic name), but according to the properties of the content itself. See "Subscription" and "Publish-subscribe model".

Content-centric A network paradigm in which the network directly provides users with content, and is aware of the content it transports, (unlike networks that limit themselves to providing communication channels between hosts).

CONVERGENCE Applications level (CoApp) The level of the CONVERGENCE architecture that establishes the interaction with CONVERGENCE users. The Applications Level interacts with the other CONVERGENCE levels on behalf of the user.

CONVERGENCE Computing Platform level (CoComp) The Computing Platform level provides content-centric networking (CoNet), secure handling (CoSec) of resources within CONVERGENCE and computing resources of peers and nodes.

CONVERGENCE Core Ontology (CCO) A semantic representation of the CoReST taxonomy. See "CONVERGENCE Resource Semantic Type (CoReST)".

CONVERGENCE Device A combination of hardware and software or a software instance that allows a user to access Convergence functionalities

CONVERGENCE Engine A collection of technologies assembled to deliver specific functionality and made available to Applications and to other Engines via an API

CONVERGENCE Middleware level (CoMid) The level of the CONVERGENCE architecture that provides the means to handle VDIs and their components.

CONVERGENCE Network (CoNet) The Content Centric component of the CONVERGENCE Computing Platform level. The CoNet provides access to named-resources on a public or private network infrastructure.

CONVERGENCE node A CONVERGENCE device that implements CoNet functionality and/or CoSec functionality.

CONVERGENCE peer A CONVERGENCE device that implements CoApp, CoMid, and CoComp (CoNet and CoSec) functionality.

CONVERGENCE Resource Semantic Type (CoReST) A list of concepts or terms that makes it possible to categorize a resource, establishing a connection with the resource's semantic metadata.

CONVERGENCE Security element (CoSec) A component of the CONVER-GENCE Computing Platform level implementing basic security functionality such as storage of private keys, basic cryptography, etc.

CONVERGENCE System A system consisting of a set of interconnected devices—peers and nodes—connected to each other built by using the technologies specified or adopted by the CONVERGENCE specification. See "Node" and "Peer".

Dec_Key_Unwrap Key Unwrapping and Content Decryption

DIDL Digital Item Description Language

Digital forgetting A CONVERGENCE system functionality ensuring that VDIs do not remain accessible for indefinite periods of time, when this is not the intention of the user.

Digital Item (DI) A structured digital object with a standard representation, identification and metadata. A DI consists of resource, resource and context related metadata, and structure. The structure is given by a Digital Item Declaration (DID) that links resource and metadata.

Domain ontology An ontology, dedicated to a specific domain of knowledge or application, e.g. the W3C Time Ontology and the GeoNames ontology.

Elementary Service (ES) The most basic service functionality offered by the CoMid.

Enc_Key_Wrap Encryption and Key Wrapping.

Entity An object, e.g. VDIs, resources, devices, events, group, licenses/contracts, services and users, that an Elementary Service can act upon or with which it can interact.

Expiry date The last date on which a VDI is accessible by a user of the CONVERGENCE System.

Fractal A semantically defined virtual cluster of CONVERGENCE peers.

Group_Sigc Group Signature.

ICN Information Centric Network.

Identifier A unique signifier assigned to a VDI or components of a VDI.

Integration Architecture An implementation of CoNet designed to integrate CoNet functionality in the IP protocol by means of a novel IPv4 option or by means of an IPv6 extension header, making IP content-aware. See "Clean-state Architecture", "Overlay Architecture" and "Parallel Architecture".

IP Identity Provider.

License A machine-readable expression of Operations that may be executed by a Principal.

Local named resource A named-resource made available to CONVERGENCE users through a local device, permanently connected to the network. Users have two options to make named-resources available to other users: (1) store the resource in a device, with a permanent connection to the network; (2) use a hosting service. In the event she chooses the former option, the resource is referred to as a local named-resource.

Metadata Data describing a resource, including but not limited to provenance, classification, expiry date etc.

MPEG eXtensible Middleware (MXM) A standard Middleware specifying a set of Application Programming Interfaces (APIs) so that MXM Applications executing on an MXM Device can access the standard multimedia technologies contained in the Middleware as MXM Engines.

MPEG-M An emerging ISO/IEC standard that includes the previous MXM standard.

Multi-homing In the context of IP networks, the configuration of multiple network interfaces or IP addresses on a single computer.

Named resource A CoNet resource that can be identified by means of a name. Named-resources may be either data (in the following referred to as "named-data") or service-access-points ("named-service-access-points").

Named service access point A kind of named-resource, consisting of a service access point identified by a name. A named-service-access-point is a network endpoint identified by its name rather than by the Internet port numbering mechanism.

Named-data A named-resource consisting of data.

Network Identifier (NID) An identifier identifying a named resource in the CONVERGENCE Network. If the named resource is a VDI or an identified VDI component, its NID may be derived from the Identifier (see "Identifier").

Overlay architecture An implementation of CoNet as an overlay over IP. See "Clean-state Architecture", "Integration Architecture" and "Parallel Architecture".

Parallel architecture An implementation of CoNet as a new networking layer that can be used in parallel to IP. See "Clean-state Architecture", "Integration Architecture" and "Overlay Architecture".

PKI Public Key Infrastructure.

Policy routing In the context of IP networks, a collection of tools for forwarding and routing data packets based on policies defined by network administrators.

Principal (CoNet) The user who is granted the right to use a CoNet Principal Identifier for naming its named resources. For example, the principal could be the provider of a service, the publisher or the author of a book, the controller of a traffic lights infrastructure, or, in general, the publisher of a VDI. A Principal may have several Principal Identifiers in the CoNet.

Principal (Rights Expression Language) The User to whom Permissions are Granted in a License.

Principal Identifier (CoNet) The Principal identifier is a string that is used in the Network Identifiers (NID) of a CoNet resource, when the NID has the form: NID = <namespace ID, hash (Principal Identifier), hash (Label)> In this approach, hash (Principal Identifier) must be unique in the namespace ID, and Label is a string chosen by the principal in such a way that hash(Label) is unique for in the context of the Principal Identifier.

Publish The act of informing an identified subset of users of the CONVERGENCE System that a VDI is available.

Publisher A user of CONVERGENCE who performs the act of publishing.

Publish-subscribe model CONVERGENCE uses a content-based approach for the publish-subscribe model, in which notifications about VDIs are delivered to a subscriber only if the metadata/content of those VDIs match constraints defined by the subscriber in his Subscription VDI.

Real World Object A physical object that may be referenced by a VDI.

REL Rights Expression Language.

Resource A virtual or physical object or service referenced by a VDI, e.g. media, Real World Objects, persons, internet services.

Scope (in the context of routing) In the context of advertising and routing, the geographical or administrative domain on which a network function operates (e.g. a well-defined section of the network—a campus, a shopping mall, an airport—, or to a subset of nodes that receives advertisements from a service provider).

Search The act through which a user requests a list of VDIs meeting a set of search criteria (e.g. specific key value pairs in the metadata, key words, free text etc.).

Serv_Auth Server Authentication without Smart Card.

Service Level Agreement (SLA) An agreement between a service provider and another user or another service provider of CONVERGENCE to provide the latter with a service whose quality matches parameters defined in the agreement.

Sig Signature.

Smart_Card Role_Auth_SC Role Authentication towards Smart Card.

SP Service Provider.

Subscribe The act whereby a user requests notification every time another user publishes or updates a VDI that satisfies the subscription criteria defined by the former user (key value pairs in the metadata, free text, key words etc.).

Subscriber A user of CONVERGENCE who performs the act of subscribing.

Timestamp A machine-readable representation of a date and time.

Tool Software providing a specific functionality that can be re-used in several applications.

Trials Organized tests of the CONVERGENCE System in specific business scenarios.

Un-named-data A data resource with no NID.

Us_Reg_IP User Registration to Identity Provider.

Us_Reg_SP User Registration to Service Provider.

User Any person or legal entity in a Value-Chain connecting (and including) Creator and End-User possibly via other Users.

User (in OSI sense) In a layered architecture, the term is used to identify an entity exploiting the service provided by a layer (e.g. CoNet user).

User ontology An ontology created by CONVERGENCE users when publishing or subscribing to a VDI.

User Profile A description of the attributes and credentials of a user of the CONVERGENCE System.

Versatile Digital Item (VDI) A structured, hierarchically organized, digital object containing one or more resources and metadata, including a declaration of the parts that make up the VDI and the links between them.

Author Index

F. Almeida et al. (eds.), *Enhancing the Internet with the CONVERGENCE System*,
Signals and Communication Technology, DOI: 10.1007/978-1-4471-5373-3,
© Springer-Verlag London 2014

Subject Index

F. Almeida et al. (eds.), *Enhancing the Internet with the CONVERGENCE System,*
Signals and Communication Technology, DOI: 10.1007/978-1-4471-5373-3,
© Springer-Verlag London 2014

Printed by Books on Demand, Germany